Ben Stacy Jerrik (Ed.)

Text Sim

Ben Stacy Jerrik (Ed.)

Text Sim

Computer, Video game, Football Manager

Part Press

Imprint

Publisher:
Part Press is a trademark of
International Book Market Service Ltd., 17 Rue Meldrum, Beau Bassin, 1713-01 Mauritius
Email: info@bookmarketservice.com
Website: www.bookmarketservice.com

Published in 2012

Printed in: U.S.A., U.K., Germany. This book was not produced in Mauritius.

ISBN: 978-620-1-64483-0

Contents

Articles

Text_sim 1
Computer 4
Video_game 23
Football Manager 37
Championship Manager 42
Sports_Interactive 44
Front_Office_Football 47
Lemonade Stand 48
Government_simulation_game 49

References

Article Sources and Contributors 54
Image Sources, Licenses and Contributors 57

Text_sim

Text Sim (which stands for text simulation) refers to a type of computer or video game that focuses on using a text based element to simulate some aspect of the real world. Text sims typically focus on creating as detailed a simulation of their object as possible, and therefore, other traditional game elements are often set aside in pursuit of creating an accurate simulation experience for the user. This pursuit of accurate simulation often comes at the expense of some or most audio or graphical elements. Numerous examples of soundless and graphic-light text sims exist.

The best selling text sim of all time is the Championship Manager/Football Manager series, published by Sports Interactive games, sold in America as Worldwide Soccer Manager. This soccer text simulation is an advanced version of a text sim with a 2D graphical sim engine for soccer games.

There are several genres of text sims. The most popular may be the sports text sim. In addition to the above mentioned Football Manager, there are several text-based sim games in baseball, football, basketball, hockey, even wrestling. This genre is typically heavy is stats, and some games have been compared to spreadsheets with all of their detail to simulating the true environment of the sport. One of the flagship products of the sports text sim is Front Office Football, which is renowned for creating a very accurate experience as a general manager of a professional football team while also retaining a heavy statistical engine.

Another genre of text sims is the tycoon genre. Although early entrants in the tycoon genre were graphically based, such as Railroad Tycoon and later Roller Coaster Tycoon, a large number of tycoon clones arose. The tycoon genre focuses on an economic simulation of one or more commercial elements. Many recent budget tycoon games have more in common with text sims than the original games, often retaining graphics merely as an interface to appeal to the general audience. Examples of these games include Coffee Tycoon, where a player runs a coffee shop franchise; Hollywood Mogul, where the player creates and runs a movie studio; and Starship Tycoon where the player manages a merchant spaceship plying the trade routes in the future. One of the earliest text sims was a simplistic economics-based game, called Lemonade Stand, where the player takes on the role of a child managing his lemonade stand.

A third text sim genre is the political simulation. Political enterprises such as elections lend themselves well to the statistics-oriented text sim gaming. These text sims often focus on elections, although a few chose instead to focus on the running on a country. Due to the nature of this genre, it often relies on a graphical interface, typically a map. However, the game information is delivered in a text-based format. Examples of this genre include The Political Machine and Power Politics, where the player tries to win an election for the American Presidency.

Yet another text sim genre is the real world simulation. These text sims try to manage a real world situation of some sort in a text based engine. Often this genre will try to use statistics to figure out some element of life. A recent example of this genre is Kudos, where a player simulates the life of a young person in their early twenties trying to decide what to do with life.

Popular Text Sims

Sports Sims:

- Bowl Bound College Football
- Baseball Mogul
- Deeproute
- Diamond Mind Baseball
- Draft Day Sports: Pro Basketball
- Draft Day Sports: College Basketball
- Eastside Hockey Manager series
- Extreme Warfare
- Fastbreak Basketball
- Fastbreak College Basketball
- Football Manager
- Football Mogul
- Footy Fanatic
- Front Office Football
- Hattrick
- Hooves of Thunder
- Inside the Park Baseball
- Jump Shot Basketball
- QuarterPole Plus
- QOOTY
- Out of the Park Baseball
- Professional Football Simulator
- PureSim
- Sick as a Parrot
- Front Office Football: The College Years
- Title Bout Championship Boxing
- Total College Basketball
- Total Extreme Wrestling
- Total Pro Basketball
- Total Pro Football
- Total Pro Golf
- Tournament Dreams College Basketball

Tycoon/Business Sims:

- Chart Wars 3
- Coffee Tycoon
- Hollywood Mogul
- Lemonade Stand
- Lemonade Tycoon
- Music Wars
- Music Maven
- Railroad Tycoon
- Starship Tycoon
- TV Manager
- Wall Street Raider

Political Sims:

- Diplomacy
- The Political Machine
- Power Politics
- President 2000

Computer

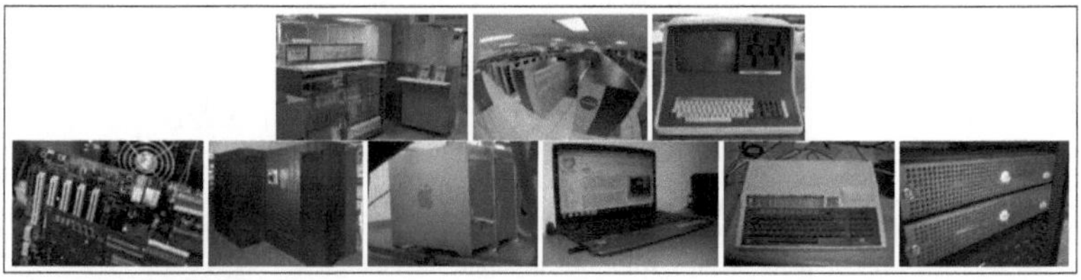

A **computer** is a general purpose device which can be programmed to carry out a finite set of arithmetic or logical operations. Since a sequence of operations can be readily changed, the computer can solve more than one kind of problem. The essential point of a computer is to implement an *idea*, the terms of which are satisfied by Alan Turing's Universal Turing machine.

Conventionally, a computer consists of at least one processing element and some form of memory. The processing element carries out arithmetic and logic operations, and a sequencing and control unit that can change the order of operations based on stored information. Peripheral devices allow information to be retrieved from an external source, and the result of operations saved.

A computer's processing unit executes a series of instructions that make it read, manipulate and then store data. Conditional instructions change the sequence of instructions as a function of the current state of the machine or its environment.

In order to interact with such a machine, programmers and engineers developed the concept of a user interface in order to accept input from humans and return results for human consumption.

The first electronic digital computers were developed between 1940 and 1945 in the United Kingdom and United States. Originally, they were the size of a large room, consuming as much power as several hundred modern personal computers (PCs).[1] In this era mechanical analog computers were used for military applications.

Modern computers based on integrated circuits are millions to billions of times more capable than the early machines, and occupy a fraction of the space.[2] Simple computers are small enough to fit into mobile devices, and mobile computers can be powered by small batteries. Personal computers in their various forms are icons of the Information Age and are what most people think of as "computers". However, the embedded computers found in many devices from mp3 players to fighter aircraft and from toys to industrial robots are the most numerous.

History of computing

The first use of the word "computer" was recorded in 1613, referring to a person who carried out calculations, or computations, and the word continued with the same meaning until the middle of the 20th century. From the end of the 19th century the word began to take on its more familiar meaning, a machine that carries out computations.[3]

The Jacquard loom, on display at the Museum of Science and Industry in Manchester, England, was one of the first programmable devices.

Limited-function early computers

The history of the modern computer begins with two separate technologies, automated calculation and programmability, but no single device can be identified as the earliest computer, partly because of the inconsistent application of that term. A few devices are worth mentioning though, like some mechanical aids to computing, which were very successful and survived for centuries until the advent of the electronic calculator, like the Sumerian abacus, designed around 2500 BC[4] of which a descendant won a speed competition against a modern desk calculating machine in Japan in 1946,[5] the slide rules, invented in the 1620s, which were carried on five Apollo space missions, including to the moon[6] and arguably the astrolabe and the Antikythera mechanism, an ancient astronomical computer built by the Greeks around 80 BC.[7] The Greek mathematician Hero of Alexandria (c. 10–70 AD) built a mechanical theater which performed a play lasting 10 minutes and was operated by a complex system of ropes and drums that might be considered to be a means of deciding which parts of the mechanism performed which actions and when.[8] This is the essence of programmability.

Around the end of the 10th century, the French monk Gerbert d'Aurillac brought back from Spain the drawings of a machine invented by the Moors that answered either Yes or No to the questions it was asked.[9] Again in the 13th century, the monks Albertus Magnus and Roger Bacon built talking androids without any further development (Albertus Magnus complained that he had wasted forty years of his life when Thomas Aquinas, terrified by his machine, destroyed it).[10]

In 1642, the Renaissance saw the invention of the mechanical calculator,[9] a device that could perform all four arithmetic operations without relying on human intelligence.[11] The mechanical calculator was at the root of the development of computers in two separate ways. Initially, it was in trying to develop more powerful and more flexible calculators[12] that the computer was first theorized by Charles Babbage[13] [14] and then developed.[15] Secondly, development of a low-cost electronic calculator, successor to the mechanical calculator, resulted in the development by Intel[16] of the first commercially available microprocessor integrated circuit.

First general-purpose computers

In 1801, Joseph Marie Jacquard made an improvement to the textile loom by introducing a series of punched paper cards as a template which allowed his loom to weave intricate patterns automatically. The resulting Jacquard loom was an important step in the development of computers because the use of punched cards to define woven patterns can be viewed as an early, albeit limited, form of programmability.

The Most Famous Image in the Early History of Computing From cave paintings to the internet HistoryofScience.comThis portrait of Jacquard was woven in silk on a Jacquard loom and required 24,000 punched cards to create (1839). It was only produced to order. Charles Babbage owned one of these portraits ; it inspired him in using perforated cards in his analytical engineSee: Anthony Hyman, ed., Science and Reform: Selected Works of Charles Babbage (Cambridge, England: Cambridge University Press, 1989), page 298. It is in the collection of the Science Museum in London, England. (Delve (2007), page 99.)

It was the fusion of automatic calculation with programmability that produced the first recognizable computers. In 1837, Charles Babbage was the first to conceptualize and design a fully programmable mechanical computer, his analytical engine.[19] Limited finances and Babbage's inability to resist tinkering with the design meant that the device was never completed—nevertheless his son, Henry Babbage, completed a simplified version of the analytical engine's computing unit (the *mill*) in 1888. He gave a successful demonstration of its use in computing tables in 1906. This machine was given to the Science museum in South Kensington in 1910.

In the late 1880s, Herman Hollerith invented the recording of data on a machine-readable medium. Earlier uses of machine-readable media had been for control, not data. "After some initial trials with paper tape, he settled on punched cards ..."[20] To process these punched cards he invented the tabulator, and the keypunch machines. These three inventions were the foundation of the modern information processing industry. Large-scale automated data processing of punched cards was performed for the 1890 United States Census by Hollerith's company, which later became the core of IBM. By the end of the 19th century a number of ideas and technologies, that would later prove useful in the realization of practical computers, had begun to appear: Boolean algebra, the vacuum tube (thermionic valve), punched cards and tape, and the teleprinter.

During the first half of the 20th century, many scientific computing needs were met by increasingly sophisticated analog computers, which used a direct mechanical or electrical model of the problem as a basis for computation. However, these were not programmable and generally lacked the versatility and accuracy of modern digital computers.

Alan Turing is widely regarded as the father of modern computer science. In 1936 Turing provided an influential formalisation of the concept of the algorithm and computation with the Turing machine, providing a blueprint for the electronic digital computer.[21] Of his role in the creation of the modern computer, *Time* magazine in naming Turing one of the 100 most influential people of the 20th century, states: "The fact remains that everyone who taps at a keyboard, opening a spreadsheet or a word-processing program, is working on an incarnation of a Turing machine".[21]

The Zuse Z3, 1941, considered the world's first working programmable, fully automatic computing machine.

The Atanasoff–Berry Computer (ABC) was the world's first electronic digital computer, albeit not programmable.[22] Atanasoff is considered to be one of the fathers of the computer.[23] Conceived in 1937 by Iowa State College physics professor John Atanasoff, and built with the assistance of graduate student Clifford Berry,[24] the machine was not programmable, being designed only to solve systems of linear equations. The computer did employ parallel computation. A 1973 court ruling in a patent dispute found that the patent for the 1946 ENIAC computer derived from the Atanasoff–Berry Computer.

The ENIAC, which became operational in 1946, is considered to be the first general-purpose electronic computer.

The first program-controlled computer was invented by Konrad Zuse, who built the Z3, an electromechanical computing machine, in 1941.[25] The first programmable electronic computer was the Colossus, built in 1943 by Tommy Flowers.

George Stibitz is internationally recognized as a father of the modern digital computer. While working at Bell Labs in November 1937, Stibitz invented and built a relay-based calculator he dubbed the "Model K" (for "kitchen table", on which he had assembled it), which was the first to use binary circuits to perform an arithmetic operation. Later models added greater sophistication including complex arithmetic and programmability.[26]

EDSAC was one of the first computers to implement the stored-program (von Neumann) architecture.

A succession of steadily more powerful and flexible computing devices were constructed in the 1930s and 1940s, gradually adding the key features that are seen in modern computers. The use of digital electronics (largely invented by Claude Shannon in 1937) and more flexible programmability were vitally important steps, but defining one point along this road as "the first digital electronic computer" is difficult.[Shannon 1940] Notable achievements include:

- Konrad Zuse's electromechanical "Z machines". The Z3 (1941) was the first working machine featuring binary arithmetic, including floating point arithmetic and a measure of programmability. In 1998 the Z3 was proved to be Turing complete, therefore being the world's first operational computer.[27]
- The non-programmable Atanasoff–Berry Computer (commenced in 1937, completed in 1941) which used vacuum tube based computation, binary numbers, and regenerative capacitor memory. The use of regenerative

memory allowed it to be much more compact than its peers (being approximately the size of a large desk or workbench), since intermediate results could be stored and then fed back into the same set of computation elements.

- The secret British Colossus computers (1943),[28] which had limited programmability but demonstrated that a device using thousands of tubes could be reasonably reliable and electronically reprogrammable. It was used for breaking German wartime codes.
- The Harvard Mark I (1944), a large-scale electromechanical computer with limited programmability.[29]
- The U.S. Army's Ballistic Research Laboratory ENIAC (1946), which used decimal arithmetic and is sometimes called the first general purpose electronic computer (since Konrad Zuse's Z3 of 1941 used electromagnets instead of electronics). Initially, however, ENIAC had an inflexible architecture which essentially required rewiring to change its programming.

Stored-program architecture

Several developers of ENIAC, recognizing its flaws, came up with a far more flexible and elegant design, which came to be known as the "stored-program architecture" or von Neumann architecture. This design was first formally described by John von Neumann in the paper *First Draft of a Report on the EDVAC*, distributed in 1945. A number of projects to develop computers based on the stored-program architecture commenced around this time, the first of which was completed in 1948 at the University of Manchester in England, the Manchester Small-Scale Experimental Machine (SSEM or "Baby"). The Electronic Delay Storage Automatic Calculator (EDSAC), completed a year after the SSEM at Cambridge University, was the first practical, non-experimental implementation of the stored-program design and was put to use immediately for research work at the university. Shortly thereafter, the machine originally described by von Neumann's paper—EDVAC—was completed but did not see full-time use for an additional two years.

Nearly all modern computers implement some form of the stored-program architecture, making it the single trait by which the word "computer" is now defined. While the technologies used in computers have changed dramatically since the first electronic, general-purpose computers of the 1940s, most still use the von Neumann architecture.

Beginning in the 1950s, Soviet scientists Sergei Sobolev and Nikolay Brusentsov conducted research on ternary computers, devices that operated on a base three numbering system of −1, 0, and 1 rather than the conventional binary numbering system upon which most computers are based. They designed the Setun, a functional ternary computer, at Moscow State University. The device was put into limited production in the Soviet Union, but supplanted by the more common binary architecture.

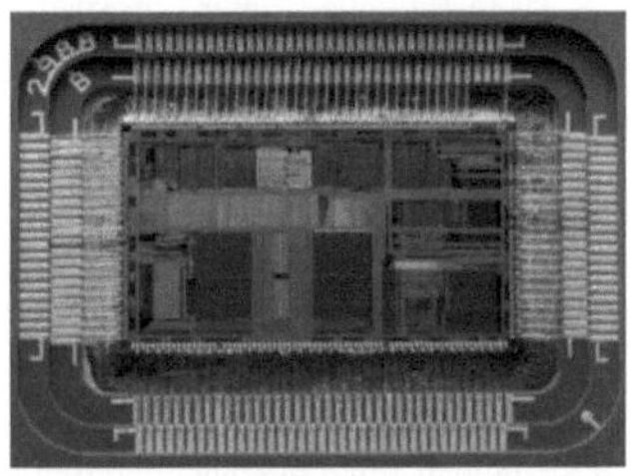

Die of an Intel 80486DX2 microprocessor (actual size: 12×6.75 mm) in its packaging.

Semiconductors and microprocessors

Computers using vacuum tubes as their electronic elements were in use throughout the 1950s, but by the 1960s had been largely replaced by semiconductor transistor-based machines, which were smaller, faster, cheaper to produce, required less power, and were more reliable. The first transistorised computer was demonstrated at the University of Manchester in 1953.[30] In the 1970s, integrated circuit technology and the subsequent creation of microprocessors, such as the Intel 4004, further decreased size and cost and further increased speed and reliability of computers. By the late 1970s, many products such as video recorders contained dedicated computers called microcontrollers, and they started to appear as a replacement to mechanical controls in domestic appliances such as washing machines. The 1980s witnessed home computers and the now ubiquitous personal computer. With the evolution of the Internet, personal computers are becoming as common as the television and the telephone in the household.

Modern smartphones are fully programmable computers in their own right, and as of 2009 may well be the most common form of such computers in existence.

Programs

The defining feature of modern computers which distinguishes them from all other machines is that they can be programmed. That is to say that some type of instructions (the program) can be given to the computer, and it will process them. While some computers may have strange concepts "instructions" and "output" (see quantum computing), modern computers based on the von Neumann architecture often have machine code in the form of an imperative programming language.

Replica of the Small-Scale Experimental Machine (SSEM), the world's first stored-program computer, at the Museum of Science and Industry in Manchester, England

In practical terms, a computer program may be just a few instructions or extend to many millions of instructions, as do the programs for word processors and web browsers for example. A typical modern computer can execute billions of instructions per second (gigaflops) and rarely makes a mistake over many years of operation. Large computer programs consisting of several million instructions may take teams of programmers years to write, and due to the complexity of the task almost certainly contain errors.

Stored program architecture

This section applies to most common RAM machine-based computers.

In most cases, computer instructions are simple: add one number to another, move some data from one location to another, send a message to some external device, etc. These instructions are read from the computer's memory and are generally carried out (executed) in the order they were given. However, there are usually specialized instructions to tell the computer to jump ahead or backwards to some other place in the program and to carry on executing from there. These are called "jump" instructions (or branches). Furthermore, jump instructions may be made to happen conditionally so that different sequences of instructions may be used depending on the result of some previous calculation or some external event. Many computers directly support subroutines by providing a type of jump that "remembers" the location it jumped from and another instruction to return to the instruction following that jump instruction.

Program execution might be likened to reading a book. While a person will normally read each word and line in sequence, they may at times jump back to an earlier place in the text or skip sections that are not of interest. Similarly, a computer may sometimes go back and repeat the instructions in some section of the program over and over again until some internal condition is met. This is called the flow of control within the program and it is what allows the computer to perform tasks repeatedly without human intervention.

Comparatively, a person using a pocket calculator can perform a basic arithmetic operation such as adding two numbers with just a few button presses. But to add together all of the numbers from 1 to 1,000 would take thousands of button presses and a lot of time, with a near certainty of making a mistake. On the other hand, a computer may be programmed to do this with just a few simple instructions. For example:

```
      mov No. 0, sum      ; set sum to 0
      mov No. 1, num      ; set num to 1
loop: add num, sum      ; add num to sum
      add No. 1, num      ; add 1 to num
      cmp num, #1000    ; compare num to 1000
      ble loop          ; if num <= 1000, go back to 'loop'
```

```
    halt              ; end of program. stop running
```

Once told to run this program, the computer will perform the repetitive addition task without further human intervention. It will almost never make a mistake and a modern PC can complete the task in about a millionth of a second.[31]

Bugs

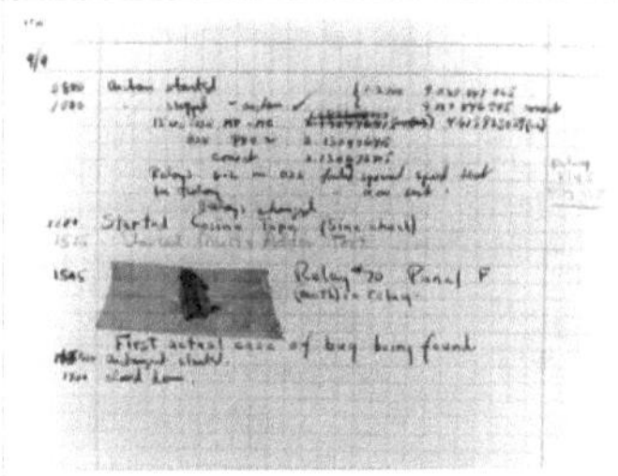

The actual first computer bug, a moth found trapped on a relay of the Harvard Mark II computer

Errors in computer programs are called "bugs". They may be benign and not affect the usefulness of the program, or have only subtle effects. But in some cases they may cause the program or the entire system to "hang" – become unresponsive to input such as mouse clicks or keystrokes – to completely fail, or to crash. Otherwise benign bugs may sometimes be harnessed for malicious intent by an unscrupulous user writing an exploit, code designed to take advantage of a bug and disrupt a computer's proper execution. Bugs are usually not the fault of the computer. Since computers merely execute the instructions they are given, bugs are nearly always the result of programmer error or an oversight made in the program's design.[32]

Rear Admiral Grace Hopper is credited for having first used the term "bugs" in computing after a dead moth was found shorting a relay in the Harvard Mark II computer in September 1947.[33]

Machine code

In most computers, individual instructions are stored as machine code with each instruction being given a unique number (its operation code or opcode for short). The command to add two numbers together would have one opcode, the command to multiply them would have a different opcode and so on. The simplest computers are able to perform any of a handful of different instructions; the more complex computers have several hundred to choose from, each with a unique numerical code. Since the computer's memory is able to store numbers, it can also store the instruction codes. This leads to the important fact that entire programs (which are just lists of these instructions) can be represented as lists of numbers and can themselves be manipulated inside the computer in the same way as numeric data. The fundamental concept of storing programs in the computer's memory alongside the data they operate on is the crux of the von Neumann, or stored program, architecture. In some cases, a computer might store some or all of its program in memory that is kept separate from the data it operates on. This is called the Harvard architecture after the Harvard Mark I computer. Modern von Neumann computers display some traits of the Harvard architecture in their designs, such as in CPU caches.

While it is possible to write computer programs as long lists of numbers (machine language) and while this technique was used with many early computers,[34] it is extremely tedious and potentially error-prone to do so in practice, especially for complicated programs. Instead, each basic instruction can be given a short name that is indicative of its function and easy to remember – a mnemonic such as ADD, SUB, MULT or JUMP. These mnemonics are collectively known as a computer's assembly language. Converting programs written in assembly language into something the computer can actually understand (machine language) is usually done by a computer program called an assembler.

Programming language

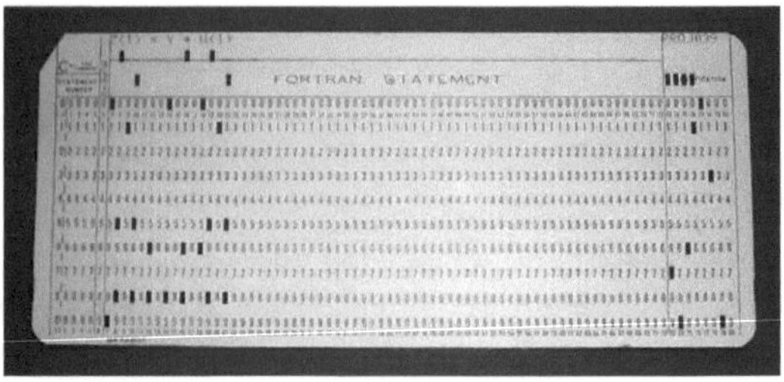

A 1970s punched card containing one line from a FORTRAN program. The card reads: "Z(1) = Y + W(1)" and is labelled "PROJ039" for identification purposes.

Programming languages provide various ways of specifying programs for computers to run. Unlike natural languages, programming languages are designed to permit no ambiguity and to be concise. They are purely written languages and are often difficult to read aloud. They are generally either translated into machine code by a compiler or an assembler before being run, or translated directly at run time by an interpreter. Sometimes programs are executed by a hybrid method of the two techniques.

Low-level languages

Machine languages and the assembly languages that represent them (collectively termed *low-level programming languages*) tend to be unique to a particular type of computer. For instance, an ARM architecture computer (such as may be found in a PDA or a hand-held videogame) cannot understand the machine language of an Intel Pentium or the AMD Athlon 64 computer that might be in a PC.[35]

Higher-level languages

Though considerably easier than in machine language, writing long programs in assembly language is often difficult and is also error prone. Therefore, most practical programs are written in more abstract high-level programming languages that are able to express the needs of the programmer more conveniently (and thereby help reduce programmer error). High level languages are usually "compiled" into machine language (or sometimes into assembly language and then into machine language) using another computer program called a compiler.[36] High level languages are less related to the workings of the target computer than assembly language, and more related to the language and structure of the problem(s) to be solved by the final program. It is therefore often possible to use different compilers to translate the same high level language program into the machine language of many different types of computer. This is part of the means by which software like video games may be made available for different computer architectures such as personal computers and various video game consoles.

Program design

Program design of small programs is relatively simple and involves the analysis of the problem, collection of inputs, using the programming constructs within languages, devising or using established procedures and algorithms, providing data for output devices and solutions to the problem as applicable. As problems become larger and more complex, features such as subprograms, modules, formal documentation, and new paradigms such as object-oriented programming are encountered. Large programs involving thousands of line of code and more require formal software methodologies. The task of developing large software systems presents a significant intellectual challenge. Producing software with an acceptably high reliability within a predictable schedule and budget has historically been difficult; the academic and professional discipline of software engineering concentrates specifically on this challenge.

Components

A general purpose computer has four main components: the arithmetic logic unit (ALU), the control unit, the memory, and the input and output devices (collectively termed I/O). These parts are interconnected by busses, often made of groups of wires.

Inside each of these parts are thousands to trillions of small electrical circuits which can be turned off or on by means of an electronic switch. Each circuit represents a bit (binary digit) of information so that when the circuit is on it represents a "1", and when off it represents a "0" (in positive logic representation). The circuits are arranged in logic gates so that one or more of the circuits may control the state of one or more of the other circuits.

The control unit, ALU, registers, and basic I/O (and often other hardware closely linked with these) are collectively known as a central processing unit (CPU). Early CPUs were composed of many separate components but since the mid-1970s CPUs have typically been constructed on a single integrated circuit called a *microprocessor*.

Control unit

The control unit (often called a control system or central controller) manages the computer's various components; it reads and interprets (decodes) the program instructions, transforming them into a series of control signals which activate other parts of the computer.[37] Control systems in advanced computers may change the order of some instructions so as to improve performance.

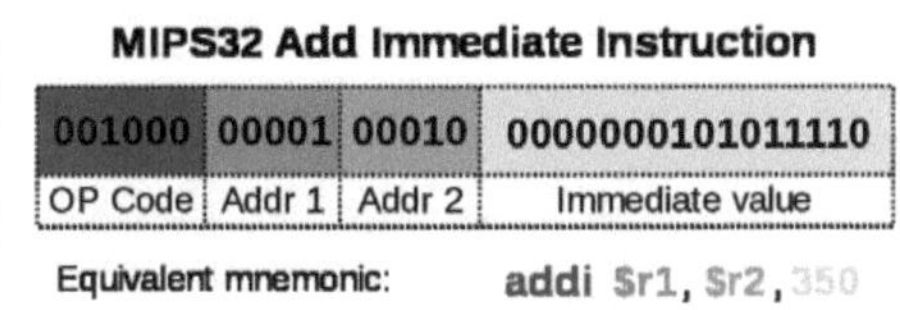

Diagram showing how a particular MIPS architecture instruction would be decoded by the control system.

A key component common to all CPUs is the program counter, a special memory cell (a register) that keeps track of which location in memory the next instruction is to be read from.[38]

The control system's function is as follows—note that this is a simplified description, and some of these steps may be performed concurrently or in a different order depending on the type of CPU:

1. Read the code for the next instruction from the cell indicated by the program counter.
2. Decode the numerical code for the instruction into a set of commands or signals for each of the other systems.
3. Increment the program counter so it points to the next instruction.
4. Read whatever data the instruction requires from cells in memory (or perhaps from an input device). The location of this required data is typically stored within the instruction code.
5. Provide the necessary data to an ALU or register.
6. If the instruction requires an ALU or specialized hardware to complete, instruct the hardware to perform the requested operation.
7. Write the result from the ALU back to a memory location or to a register or perhaps an output device.
8. Jump back to step (1).

Since the program counter is (conceptually) just another set of memory cells, it can be changed by calculations done in the ALU. Adding 100 to the program counter would cause the next instruction to be read from a place 100 locations further down the program. Instructions that modify the program counter are often known as "jumps" and allow for loops (instructions that are repeated by the computer) and often conditional instruction execution (both examples of control flow).

The sequence of operations that the control unit goes through to process an instruction is in itself like a short computer program, and indeed, in some more complex CPU designs, there is another yet smaller computer called a microsequencer, which runs a microcode program that causes all of these events to happen.

Arithmetic logic unit (ALU)

The ALU is capable of performing two classes of operations: arithmetic and logic.[39]

The set of arithmetic operations that a particular ALU supports may be limited to addition and subtraction, or might include multiplication, division, trigonometry functions such as sine, cosine, etc., and square roots. Some can only operate on whole numbers (integers) whilst others use floating point to represent real numbers, albeit with limited precision. However, any computer that is capable of performing just the simplest operations can be programmed to break down the more complex operations into simple steps that it can perform. Therefore, any computer can be programmed to perform any arithmetic operation—although it will take more time to do so if its ALU does not directly support the operation. An ALU may also compare numbers and return boolean truth values (true or false) depending on whether one is equal to, greater than or less than the other ("is 64 greater than 65?").

Logic operations involve Boolean logic: AND, OR, XOR and NOT. These can be useful for creating complicated conditional statements and processing boolean logic.

Superscalar computers may contain multiple ALUs, allowing them to process several instructions simultaneously.[40] Graphics processors and computers with SIMD and MIMD features often contain ALUs that can perform arithmetic on vectors and matrices.

Memory

Magnetic core memory was the computer memory of choice throughout the 1960s, until it was replaced by semiconductor memory.

A computer's memory can be viewed as a list of cells into which numbers can be placed or read. Each cell has a numbered "address" and can store a single number. The computer can be instructed to "put the number 123 into the cell numbered 1357" or to "add the number that is in cell 1357 to the number that is in cell 2468 and put the answer into cell 1595". The information stored in memory may represent practically anything. Letters, numbers, even computer instructions can be placed into memory with equal ease. Since the CPU does not differentiate between different types of information, it is the software's responsibility to give significance to what the memory sees as nothing but a series of numbers.

In almost all modern computers, each memory cell is set up to store binary numbers in groups of eight bits (called a byte). Each byte is able to represent 256 different numbers (2^8 = 256); either from 0 to 255 or −128 to +127. To store larger numbers, several consecutive bytes may be used (typically, two, four or eight). When negative numbers are required, they are usually stored in two's complement notation. Other arrangements are possible, but are usually not seen outside of specialized applications or historical contexts. A computer can store any kind of information in memory if it can be represented numerically. Modern computers have billions or even trillions of bytes of memory.

The CPU contains a special set of memory cells called registers that can be read and written to much more rapidly than the main memory area. There are typically between two and one hundred registers depending on the type of CPU. Registers are used for the most frequently needed data items to avoid having to access main memory every time data is needed. As data is constantly being worked on, reducing the need to access main memory (which is often slow compared to the ALU and control units) greatly increases the computer's speed.

Computer main memory comes in two principal varieties: random-access memory or RAM and read-only memory or ROM. RAM can be read and written to anytime the CPU commands it, but ROM is pre-loaded with data and software that never changes, therefore the CPU can only read from it. ROM is typically used to store the computer's initial start-up instructions. In general, the contents of RAM are erased when the power to the computer is turned off, but ROM retains its data indefinitely. In a PC, the ROM contains a specialized program called the BIOS that orchestrates loading the computer's operating system from the hard disk drive into RAM whenever the computer is turned on or reset. In embedded computers, which frequently do not have disk drives, all of the required software may be stored in ROM. Software stored in ROM is often called firmware, because it is notionally more like hardware than software. Flash memory blurs the distinction between ROM and RAM, as it retains its data when turned off but is also rewritable. It is typically much slower than conventional ROM and RAM however, so its use is restricted to applications where high speed is unnecessary.[41]

In more sophisticated computers there may be one or more RAM cache memories, which are slower than registers but faster than main memory. Generally computers with this sort of cache are designed to move frequently needed data into the cache automatically, often without the need for any intervention on the programmer's part.

Input/output (I/O)

I/O is the means by which a computer exchanges information with the outside world.[42] Devices that provide input or output to the computer are called peripherals.[43] On a typical personal computer, peripherals include input devices like the keyboard and mouse, and output devices such as the display and printer. Hard disk drives, floppy disk drives and optical disc drives serve as both input and output devices. Computer networking is another form of I/O.

Hard disk drives are common storage devices used with computers.

I/O devices are often complex computers in their own right, with their own CPU and memory. A graphics processing unit might contain fifty or more tiny computers that perform the calculations necessary to display 3D graphics. Modern desktop computers contain many smaller computers that assist the main CPU in performing I/O.

Multitasking

While a computer may be viewed as running one gigantic program stored in its main memory, in some systems it is necessary to give the appearance of running several programs simultaneously. This is achieved by multitasking i.e. having the computer switch rapidly between running each program in turn.[44]

One means by which this is done is with a special signal called an interrupt, which can periodically cause the computer to stop executing instructions where it was and do something else instead. By remembering where it was executing prior to the interrupt, the computer can return to that task later. If several programs are running "at the same time", then the interrupt generator might be causing several hundred interrupts per second, causing a program switch each time. Since modern computers typically execute instructions several orders of magnitude faster than human perception, it may appear that many programs are running at the same time even though only one is ever executing in any given instant. This method of multitasking is sometimes termed "time-sharing" since each program is allocated a "slice" of time in turn.[45]

Before the era of cheap computers, the principal use for multitasking was to allow many people to share the same computer.

Seemingly, multitasking would cause a computer that is switching between several programs to run more slowly, in direct proportion to the number of programs it is running, but most programs spend much of their time waiting for slow input/output devices to complete their tasks. If a program is waiting for the user to click on the mouse or press a key on the keyboard, then it will not take a "time slice" until the event it is waiting for has occurred. This frees up time for other programs to execute so that many programs may be run simultaneously without unacceptable speed loss.

Multiprocessing

Cray designed many supercomputers that used multiprocessing heavily.

Some computers are designed to distribute their work across several CPUs in a multiprocessing configuration, a technique once employed only in large and powerful machines such as supercomputers, mainframe computers and servers. Multiprocessor and multi-core (multiple CPUs on a single integrated circuit) personal and laptop computers are now widely available, and are being increasingly used in lower-end markets as a result.

Supercomputers in particular often have highly unique architectures that differ significantly from the basic stored-program architecture and from general purpose computers.[46] They often feature thousands of CPUs, customized high-speed interconnects, and specialized computing hardware. Such designs tend to be useful only for specialized tasks due to the large scale of program organization required to successfully utilize most of the available resources at once. Supercomputers usually see usage in large-scale simulation, graphics rendering, and cryptography applications, as well as with other so-called "embarrassingly parallel" tasks.

Networking and the Internet

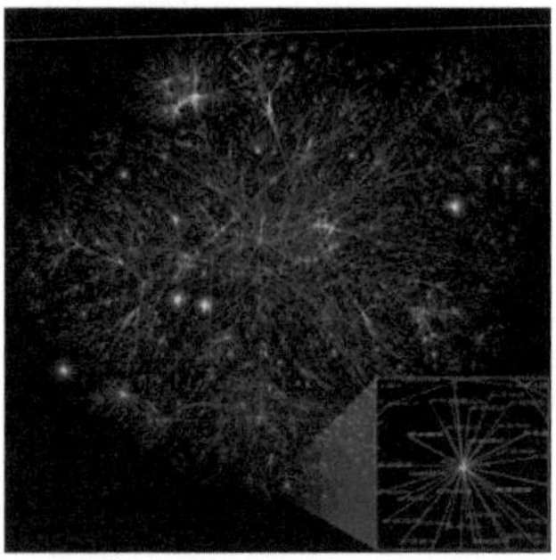

Visualization of a portion of the routes on the Internet.

Computers have been used to coordinate information between multiple locations since the 1950s. The U.S. military's SAGE system was the first large-scale example of such a system, which led to a number of special-purpose commercial systems such as Sabre.[47]

In the 1970s, computer engineers at research institutions throughout the United States began to link their computers together using telecommunications technology. The effort was funded by ARPA (now DARPA), and the computer network that resulted was called the ARPANET.[48] The technologies that made the Arpanet possible spread and evolved.

In time, the network spread beyond academic and military institutions and became known as the Internet. The emergence of networking involved a redefinition of the nature and boundaries of the computer. Computer operating systems and applications were modified to include the ability to define and access the resources of other computers on the network, such as peripheral devices, stored information, and the like, as extensions of the resources of an individual computer. Initially these facilities were available primarily to people working in high-tech environments, but in the 1990s the spread of applications like e-mail and the World Wide Web, combined with the development of cheap, fast networking technologies like Ethernet and ADSL saw computer networking become almost ubiquitous. In fact, the number of computers that are networked is growing phenomenally. A very large proportion of personal computers regularly connect to the Internet to communicate and receive information. "Wireless" networking, often utilizing mobile phone networks, has meant networking is becoming increasingly ubiquitous even in mobile computing environments.

Computer architecture paradigms

There are many types of computer architectures:

- Quantum computer vs Chemical computer
- Scalar processor vs Vector processor
- Non-Uniform Memory Access (NUMA) computers
- Register machine vs Stack machine
- Harvard architecture vs von Neumann architecture
- Cellular architecture

The quantum computer architecture holds the most promise to revolutionize computing.[49]

Logic gates are a common abstraction which can apply to most of the above digital or analog paradigms.

The ability to store and execute lists of instructions called programs makes computers extremely versatile, distinguishing them from calculators. The Church–Turing thesis is a mathematical statement of this versatility: any computer with a minimum capability (being Turing-complete) is, in principle, capable of performing the same tasks that any other computer can perform. Therefore any type of computer (netbook, supercomputer, cellular automaton, etc.) is able to perform the same computational tasks, given enough time and storage capacity.

Misconceptions

A computer does not need to be electronic, nor even have a processor, nor RAM, nor even a hard disk. While popular usage of the word "computer" is synonymous with a personal computer, the definition of a computer is literally "*A device that computes*, especially a programmable [usually] electronic machine that performs high-speed mathematical or logical operations or that assembles, stores, correlates, or otherwise processes information."[50] Any device which *processes information* qualifies as a computer, especially if the processing is purposeful.

Required technology

Historically, computers evolved from mechanical computers and eventually from vacuum tubes to transistors. However, conceptually computational systems as flexible as a personal computer can be built out of almost anything. For example, a computer can be made out of billiard balls (billiard ball computer); an oft-quoted example. More realistically, modern computers are made out of transistors made of photolithographed semiconductors.

There is active research to make computers out of many promising new types of technology, such as optical computers, DNA computers, neural computers, and quantum computers. Most computers are universal, and are able to calculate any computable function, and are limited only by their memory capacity and operating speed. However different designs of computers can give very different performance for particular problems; for example quantum computers can potentially break some modern encryption algorithms (by quantum factoring) very quickly.

Further topics

- Glossary of computers

Artificial intelligence

A computer will solve problems in exactly the way it is programmed to, without regard to efficiency, alternative solutions, possible shortcuts, or possible errors in the code. Computer programs that learn and adapt are part of the emerging field of artificial intelligence and machine learning.

Hardware

The term *hardware* covers all of those parts of a computer that are tangible objects. Circuits, displays, power supplies, cables, keyboards, printers and mice are all hardware.

History of computing hardware

First Generation (Mechanical/Electromechanical)	Calculators	Antikythera mechanism, Difference engine, Norden bombsight
	Programmable Devices	Jacquard loom, Analytical engine, Harvard Mark I, Z3
Second Generation (Vacuum Tubes)	Calculators	Atanasoff–Berry Computer, IBM 604, UNIVAC 60, UNIVAC 120
	Programmable Devices	Colossus, ENIAC, Manchester Small-Scale Experimental Machine, EDSAC, Manchester Mark 1, Ferranti Pegasus, Ferranti Mercury, CSIRAC, EDVAC, UNIVAC I, IBM 701, IBM 702, IBM 650, Z22
Third Generation (Discrete transistors and SSI, MSI, LSI Integrated circuits)	Mainframes	IBM 7090, IBM 7080, IBM System/360, BUNCH
	Minicomputer	PDP-8, PDP-11, IBM System/32, IBM System/36
Fourth Generation (VLSI integrated circuits)	Minicomputer	VAX, IBM System i
	4-bit microcomputer	Intel 4004, Intel 4040
	8-bit microcomputer	Intel 8008, Intel 8080, Motorola 6800, Motorola 6809, MOS Technology 6502, Zilog Z80
	16-bit microcomputer	Intel 8088, Zilog Z8000, WDC 65816/65802
	32-bit microcomputer	Intel 80386, Pentium, Motorola 68000, ARM architecture
	64-bit microcomputer[51]	Alpha, MIPS, PA-RISC, PowerPC, SPARC, x86-64
	Embedded computer	Intel 8048, Intel 8051
	Personal computer	Desktop computer, Home computer, Laptop computer, Personal digital assistant (PDA), Portable computer, Tablet PC, Wearable computer
Theoretical/experimental	Quantum computer, Chemical computer, DNA computing, Optical computer, Spintronics based computer	

Other hardware topics

Peripheral device (input/output)	Input	Mouse, keyboard, joystick, image scanner, webcam, graphics tablet, microphone
	Output	Monitor, printer, loudspeaker
	Both	Floppy disk drive, hard disk drive, optical disc drive, teleprinter
Computer busses	Short range	RS-232, SCSI, PCI, USB
	Long range (computer networking)	Ethernet, ATM, FDDI

Software

Software refers to parts of the computer which do not have a material form, such as programs, data, protocols, etc. When software is stored in hardware that cannot easily be modified (such as BIOS ROM in an IBM PC compatible), it is sometimes called "firmware" to indicate that it falls into an uncertain area somewhere between hardware and software.

Operating system	Unix and BSD	UNIX System V, IBM AIX, HP-UX, Solaris (SunOS), IRIX, List of BSD operating systems
	GNU/Linux	List of Linux distributions, Comparison of Linux distributions
	Microsoft Windows	Windows 95, Windows 98, Windows NT, Windows 2000, Windows Me, Windows XP, Windows Vista, Windows 7
	DOS	86-DOS (QDOS), PC-DOS, MS-DOS, DR-DOS, FreeDOS
	Mac OS	Mac OS classic, Mac OS X
	Embedded and real-time	List of embedded operating systems
	Experimental	Amoeba, Oberon/Bluebottle, Plan 9 from Bell Labs
Library	Multimedia	DirectX, OpenGL, OpenAL
	Programming library	C standard library, Standard Template Library
Data	Protocol	TCP/IP, Kermit, FTP, HTTP, SMTP
	File format	HTML, XML, JPEG, MPEG, PNG
User interface	Graphical user interface (WIMP)	Microsoft Windows, GNOME, KDE, QNX Photon, CDE, GEM, Aqua
	Text-based user interface	Command-line interface, Text user interface

Application	Office suite	Word processing, Desktop publishing, Presentation program, Database management system, Scheduling & Time management, Spreadsheet, Accounting software
	Internet Access	Browser, E-mail client, Web server, Mail transfer agent, Instant messaging
	Design and manufacturing	Computer-aided design, Computer-aided manufacturing, Plant management, Robotic manufacturing, Supply chain management
	Graphics	Raster graphics editor, Vector graphics editor, 3D modeler, Animation editor, 3D computer graphics, Video editing, Image processing
	Audio	Digital audio editor, Audio playback, Mixing, Audio synthesis, Computer music
	Software engineering	Compiler, Assembler, Interpreter, Debugger, Text editor, Integrated development environment, Software performance analysis, Revision control, Software configuration management
	Educational	Edutainment, Educational game, Serious game, Flight simulator
	Games	Strategy, Arcade, Puzzle, Simulation, First-person shooter, Platform, Massively multiplayer, Interactive fiction
	Misc	Artificial intelligence, Antivirus software, Malware scanner, Installer/Package management systems, File manager

Languages

There are thousands of different programming languages—some intended to be general purpose, others useful only for highly specialized applications.

Programming languages

Lists of programming languages	Timeline of programming languages, List of programming languages by category, Generational list of programming languages, List of programming languages, Non-English-based programming languages
Commonly used Assembly languages	ARM, MIPS, x86
Commonly used high-level programming languages	Ada, BASIC, C, C++, C#, COBOL, Fortran, Java, Lisp, Pascal, Object Pascal
Commonly used Scripting languages	Bourne script, JavaScript, Python, Ruby, PHP, Perl

Professions and organizations

As the use of computers has spread throughout society, there are an increasing number of careers involving computers.

Computer-related professions

Hardware-related	Electrical engineering, Electronic engineering, Computer engineering, Telecommunications engineering, Optical engineering, Nanoengineering
Software-related	Computer science, Desktop publishing, Human–computer interaction, Information technology, Information systems, Computational science, Software engineering, Video game industry, Web design

The need for computers to work well together and to be able to exchange information has spawned the need for many standards organizations, clubs and societies of both a formal and informal nature.

Organizations

Standards groups	ANSI, IEC, IEEE, IETF, ISO, W3C
Professional Societies	ACM, AIS, IET, IFIP, BCS
Free/Open source software groups	Free Software Foundation, Mozilla Foundation, Apache Software Foundation

See also

- Computability theory
- Computer insecurity
- Computer security
- List of computer term etymologies
- List of fictional computers
- Pulse computation

Notes

[1] In 1946, ENIAC required an estimated 174 kW. By comparison, a modern laptop computer may use around 30 W; nearly six thousand times less. "Approximate Desktop & Notebook Power Usage" (http://www.upenn.edu/computing/provider/docs/hardware/powerusage.html). University of Pennsylvania. . Retrieved 20 June 2009.

[2] Early computers such as Colossus and ENIAC were able to process between 5 and 100 operations per second. A modern "commodity" microprocessor (as of 2007) can process billions of operations per second, and many of these operations are more complicated and useful than early computer operations. "Intel Core2 Duo Mobile Processor: Features" (http://www.intel.com/cd/channel/reseller/asmo-na/eng/products/mobile/processors/core2duo_m/feature/index.htm). Intel Corporation. . Retrieved 20 June 2009.

[3] *computer,* n. (http://dictionary.oed.com/). Oxford English Dictionary (2 ed.). Oxford University Press. 1989. . Retrieved 10 April 2009.

[4] * Ifrah, Georges (2001). *The Universal History of Computing: From the Abacus to the Quantum Computer.* New York: John Wiley & Sons. ISBN 0-471-39671-0. From 2700 to 2300 BC, Georges Ifrah, pp.11

[5] Berkeley, Edmund (1949). *Giant Brains, or Machines That Think.* John Wiley & Sons. p. 19. Edmund Berkeley

[6] According to advertising on Pickett's N600 slide rule boxes. "Pickett Apollo Box Scans" (http://copland.udel.edu/~mm/sliderule/lem/). Copland.udel.edu. . Retrieved 20 February 2010.

[7] "Discovering How Greeks Computed in 100 B.C." (http://www.nytimes.com/2008/07/31/science/31computer.html?hp). *The New York Times.* 31 July 2008. . Retrieved 27 March 2010.

[8] "Heron of Alexandria" (http://www.mlahanas.de/Greeks/HeronAlexandria2.htm). . Retrieved 15 January 2008.

[9] Felt, Dorr E. (1916). *Mechanical arithmetic, or The history of the counting machine* (http://www.archive.org/details/mechanicalarithm00feltrich). Chicago: Washington Institute. p. 8. . Dorr E. Felt

[10] "Speaking machines" (http://books.google.co.uk/books?id=Xt4PAAAAYAAJ&pg=PT38&dq=the+parlour+review+january+1838&hl=en&ei=0yqzTN3kLMTHswa2wMjSDQ&sa=X&oi=book_result&ct=result&resnum=1&ved=0CCsQ6AEwAA#v=onepage&q&f=false). *The parlour review, Philadelphia* **1** (3). 20 January 1838. . Retrieved 11 October 2010.

[11] "Pascal and Leibnitz, in the seventeenth century, and Diderot at a later period, endeavored to construct a machine which might serve as a substitute for human intelligence in the combination of figures" The Gentleman's magazine, Volume 202, p.100 (http://books.google.fr/books?id=Rf0IAAAAIAAJ&pg=PA100&dq=arithmometer&as_brr=1#v=onepage&q=arithmometer&f=false)

[12] Babbage's Difference engine in 1823 and his Analytical engine in the mid-1830s

[13] "It is reasonable to inquire, therefore, whether it is possible to devise a machine which will do for mathematical computation what the automatic lathe has done for engineering. The first suggestion that such a machine could be made came more than a hundred years ago from the mathematician Charles Babbage. Babbage's ideas have only been properly appreciated in the last ten years, but we now realize that he understood clearly all the fundamental principles which are embodied in modern digital computers" *Faster than thought*, edited by B. V. Bowden, 1953, Pitman publishing corporation

[14] "...Among this extraordinary galaxy of talent Charles Babbage appears to be one of the most remarkable of all. Most of his life he spent in an entirely unsuccessful attempt to make a machine which was regarded by his contemporaries as utterly preposterous, and his efforts were regarded as futile, time-consuming and absurd. In the last decade or so we have learnt how his ideas can be embodied in a modern digital computer. He understood more about the logic of these machines than anyone else in the world had learned until after the end of the last war" Foreword, *Irascible Genius, Charles Babbage, inventor* by Maboth Moseley, 1964, London, Hutchinson

[15] In the proposal that Aiken gave IBM in 1937 while requesting funding for the Harvard Mark I we can read: "Few **calculating machines** have been designed strictly for application to scientific investigations, the notable exceptions being those of Charles Babbage and others who followed him ... After abandoning the difference engine, Babbage devoted his energy to the design and construction of an analytical engine of

far higher powers than the difference engine ... Since the time of Babbage, the development of **calculating machinery** has continued at an increasing rate." Howard Aiken, *Proposed automatic calculating machine*, reprinted in: The origins of Digital computers, Selected Papers, Edited by Brian Randell, 1973, ISBN 3-540-06169-X

[16] "Intel Museum – The 4004, Big deal then, Big deal now" (http://www.intel.com/about/companyinfo/museum/exhibits/4004/index.htm). Intel.com. . Retrieved 29 January 2012.

[17] From cave paintings to the internet (http://www.historyofscience.com/G2I/timeline/index.php?category=Computers+&+the+Human+Brain) HistoryofScience.com

[18] See: Anthony Hyman, ed., *Science and Reform: Selected Works of Charles Babbage* (Cambridge, England: Cambridge University Press, 1989), page 298. It is in the collection of the Science Museum in London, England. (Delve (2007), page 99.)

[19] The analytical engine should not be confused with Babbage's difference engine which was a non-programmable mechanical calculator.

[20] "Columbia University Computing History: Herman Hollerith" (http://www.columbia.edu/acis/history/hollerith.html). Columbia.edu. . Retrieved 11 December 2010.

[21] "Alan Turing – Time 100 People of the Century" (http://205.188.238.181/time/time100/scientist/profile/turing.html). Time Magazine. . Retrieved 13 June 2009. "The fact remains that everyone who taps at a keyboard, opening a spreadsheet or a word-processing program, is working on an incarnation of a Turing machine"

[22] "John Vincent Atanasoff and the Birth of Electronic Digital Computing" (http://www.cs.iastate.edu/jva/jva-archive.shtml). Cs.iastate.edu. . Retrieved 29 January 2012.

[23] "John Vincent Atanasoff – the father of the computer" (http://www.columbia.edu/~td2177/JVAtanasoff/JVAtanasoff.html). Columbia.edu. . Retrieved 29 January 2012.

[24] "Atanasoff-Berry Computer" (http://energysciencenews.com/phpBB3/viewtopic.php?f=1&t=98&p=264#p264). . Retrieved 20 November 2010.

[25] "Spiegel: The inventor of the computer's biography was published" (http://www.spiegel.de/netzwelt/gadgets/0,1518,651776,00.html). *Der Spiegel*. 28 September 2009. . Retrieved 11 December 2010.

[26] "Inventor Profile: George R. Stibitz" (http://www.invent.org/hall_of_fame/140.html). National Inventors Hall of Fame Foundation, Inc.. .

[27] Rojas, R. (1998). "How to make Zuse's Z3 a universal computer". *IEEE Annals of the History of Computing* **20** (3): 51–54. doi:10.1109/85.707574.

[28] B. Jack Copeland, ed., Colossus: The Secrets of Bletchley Park's Codebreaking Computers, Oxford University Press, 2006

[29] *"Robot Mathematician Knows All The Answers", October 1944, Popular Science* (http://books.google.com/books?id=PyEDAAAAMBAJ&pg=PA86&dq=motor+gun+boat&hl=en&ei=LxTqTMfGI4-bnwfEyNiWDQ&sa=X&oi=book_result&ct=result&resnum=6&ved=0CEIQ6AEwBQ#v=onepage&q=motor gun boat&f=true). Google Books. . Retrieved 11 December 2010.

[30] Lavington 1998, p. 37

[31] This program was written similarly to those for the PDP-11 minicomputer and shows some typical things a computer can do. All the text after the semicolons are comments for the benefit of human readers. These have no significance to the computer and are ignored. (Digital Equipment Corporation 1972)

[32] It is not universally true that bugs are solely due to programmer oversight. Computer hardware may fail or may itself have a fundamental problem that produces unexpected results in certain situations. For instance, the Pentium FDIV bug caused some Intel microprocessors in the early 1990s to produce inaccurate results for certain floating point division operations. This was caused by a flaw in the microprocessor design and resulted in a partial recall of the affected devices.

[33] Taylor, Alexander L., III (16 April 1984). "The Wizard Inside the Machine" (http://www.time.com/time/printout/0,8816,954266,00.html). *TIME*. . Retrieved 17 February 2007.

[34] Even some later computers were commonly programmed directly in machine code. Some minicomputers like the DEC PDP-8 could be programmed directly from a panel of switches. However, this method was usually used only as part of the booting process. Most modern computers boot entirely automatically by reading a boot program from some non-volatile memory.

[35] However, there is sometimes some form of machine language compatibility between different computers. An x86-64 compatible microprocessor like the AMD Athlon 64 is able to run most of the same programs that an Intel Core 2 microprocessor can, as well as programs designed for earlier microprocessors like the Intel Pentiums and Intel 80486. This contrasts with very early commercial computers, which were often one-of-a-kind and totally incompatible with other computers.

[36] High level languages are also often interpreted rather than compiled. Interpreted languages are translated into machine code on the fly, while running, by another program called an interpreter.

[37] The control unit's role in interpreting instructions has varied somewhat in the past. Although the control unit is solely responsible for instruction interpretation in most modern computers, this is not always the case. Many computers include some instructions that may only be partially interpreted by the control system and partially interpreted by another device. This is especially the case with specialized computing hardware that may be partially self-contained. For example, EDVAC, one of the earliest stored-program computers, used a central control unit that only interpreted four instructions. All of the arithmetic-related instructions were passed on to its arithmetic unit and further decoded there.

[38] Instructions often occupy more than one memory address, therefore the program counter usually increases by the number of memory locations required to store one instruction.

[39] David J. Eck (2000). *The Most Complex Machine: A Survey of Computers and Computing*. A K Peters, Ltd.. p. 54. ISBN 978-1-56881-128-4.

[40] Erricos John Kontoghiorghes (2006). *Handbook of Parallel Computing and Statistics*. CRC Press. p. 45. ISBN 978-0-8247-4067-2.

[41] Flash memory also may only be rewritten a limited number of times before wearing out, making it less useful for heavy random access usage. (Verma & Mielke 1988)

[42] Donald Eadie (1968). *Introduction to the Basic Computer*. Prentice-Hall. p. 12.

[43] Arpad Barna; Dan I. Porat (1976). *Introduction to Microcomputers and the Microprocessors*. Wiley. p. 85. ISBN 978-0-471-05051-3.

[44] Jerry Peek; Grace Todino, John Strang (2002). *Learning the UNIX Operating System: A Concise Guide for the New User*. O'Reilly. p. 130. ISBN 978-0-596-00261-9.

[45] Gillian M. Davis (2002). *Noise Reduction in Speech Applications*. CRC Press. p. 111. ISBN 978-0-8493-0949-6.

[46] However, it is also very common to construct supercomputers out of many pieces of cheap commodity hardware; usually individual computers connected by networks. These so-called computer clusters can often provide supercomputer performance at a much lower cost than customized designs. While custom architectures are still used for most of the most powerful supercomputers, there has been a proliferation of cluster computers in recent years. (TOP500 2006)

[47] Agatha C. Hughes (2000). *Systems, Experts, and Computers*. MIT Press. p. 161. ISBN 978-0-262-08285-3. "The experience of SAGE helped make possible the first truly large-scale commercial real-time network: the SABRE computerized airline reservations system..."

[48] "A Brief History of the Internet" (http://www.isoc.org/internet/history/brief.shtml). Internet Society. . Retrieved 20 September 2008.

[49] "Computer architecture: fundamentals and principles of computer design" (http://books.google.com/books?id=ZWaUurOwMPQC&q=quantum+computers&dq=insufficient+address+computer+architecture&source=gbs_word_cloud_r&cad=3#v=snippet&q=quantum computers&f=false) by Joseph D. Dumas 2006. page 340.

[50] "Definition of computer" (http://thefreedictionary.com/computer). Thefreedictionary.com. . Retrieved 29 January 2012.

[51] Most major 64-bit instruction set architectures are extensions of earlier designs. All of the architectures listed in this table, except for Alpha, existed in 32-bit forms before their 64-bit incarnations were introduced.

References

- [a] Kempf, Karl (1961). *Historical Monograph: Electronic Computers Within the Ordnance Corps* (http://ed-thelen.org/comp-hist/U-S-Ord-61.html). Aberdeen Proving Ground (United States Army).
- [a] Phillips, Tony (2000). "The Antikythera Mechanism I" (http://www.math.sunysb.edu/~tony/whatsnew/column/antikytheraI-0400/kyth1.html). American Mathematical Society. Retrieved 5 April 2006.
- [a] Shannon, Claude Elwood (1940). *A symbolic analysis of relay and switching circuits* (http://hdl.handle.net/1721.1/11173). Massachusetts Institute of Technology.
- Digital Equipment Corporation (1972) (PDF). *PDP-11/40 Processor Handbook* (http://bitsavers.vt100.net/dec/www.computer.museum.uq.edu.au_mirror/D-09-30_PDP11-40_Processor_Handbook.pdf). Maynard, MA: Digital Equipment Corporation.
- Verma, G.; Mielke, N. (1988). *Reliability performance of ETOX based flash memories*. IEEE International Reliability Physics Symposium.
- Meuer, Hans; Strohmaier, Erich; Simon, Horst; Dongarra, Jack (13 November 2006). "Architectures Share Over Time" (http://www.top500.org/lists/2006/11/overtime/Architectures). TOP500. Retrieved 27 November 2006.
- Lavington, Simon (1998). *A History of Manchester Computers* (2 ed.). Swindon: The British Computer Society. ISBN 978-0-902505-01-8.
- Stokes, Jon (2007). *Inside the Machine: An Illustrated Introduction to Microprocessors and Computer Architecture*. San Francisco: No Starch Press. ISBN 978-1-59327-104-6.
- Felt, Dorr E. (1916). *Mechanical arithmetic, or The history of the counting machine* (http://www.archive.org/details/mechanicalarithm00feltrich). Chicago: Washington Institute.
- Ifrah, Georges (2001). *The Universal History of Computing: From the Abacus to the Quantum Computer*. New York: John Wiley & Sons. ISBN 0-471-39671-0.
- Berkeley, Edmund (1949). *Giant Brains, or Machines That Think*. John Wiley & Sons.

External links

- A Brief History of Computing (http://www.life.com/image/first/in-gallery/48681/click-a-brief-history-of-computing#index/0) – slideshow by *Life magazine*

Video_game

A **video game** is an electronic game that involves human interaction with a user interface to generate visual feedback on a video device. The word *video* in *video game* traditionally referred to a raster display device,[1] but following popularization of the term "video game", it now implies any type of display device. The electronic systems used to play video games are known as platforms; examples of these are personal computers and video game consoles. These platforms range from large mainframe computers to small handheld devices. Specialized video games such as arcade games, while previously common, have gradually declined in use. Video games have gone on to become an art form and industry.

The input device used to manipulate video games is called a game controller, and varies across platforms. For example, a controller might consist of only a button and a joystick, while another may feature a dozen buttons and one or more joysticks. Early personal computer games often needed a keyboard for gameplay, or more commonly, required the user to buy a separate joystick with at least one button.[2] Many modern computer games allow or require the player to use a keyboard and a mouse simultaneously. A few of the most common game controllers are gamepads, mice, keyboards, and joysticks.

Video games typically use additional means of providing interactivity and information to the player. Audio is almost universal, using sound reproduction devices, such as speakers and headphones. Other feedback may come via haptic peripherals, such as vibration or force feedback, with vibration sometimes used to simulate force feedback.

In the early days of cartridge consoles, they were sometimes called **TV games**.

History

Early games used interactive electronic devices with various display formats. The earliest example is from 1947—a "Cathode ray tube Amusement Device" was filed for a patent on January 25, 1947, by Thomas T. Goldsmith Jr. and Estle Ray Mann, and issued on December 14, 1948, as U.S. Patent 2455992.[3]

Inspired by radar display tech, it consisted of an analog device that allowed a user to control a vector-drawn dot on the screen to simulate a missile being fired at targets, which were drawings fixed to the screen.[4]

Other early examples include:

- The NIMROD computer at the 1951 Festival of Britain
- *OXO* a tic-tac-toe Computer game by Alexander S. Douglas for the EDSAC in 1952
- *Tennis for Two*, an interactive game engineered by William Higinbotham in 1958
- *Spacewar!*, written by MIT students Martin Graetz, Steve Russell, and Wayne Wiitanen's on a DEC PDP-1 computer in 1961.

Each game used different means of display: NIMROD used a panel of lights to play the game of Nim,[5] OXO used a graphical display to play tic-tac-toe [6] *Tennis for Two* used an oscilloscope to display a side view of a tennis court,[4] and *Spacewar!* used the DEC PDP-1's vector display to have two spaceships battle each other.[7]

Nolan Bushnell at the Game Developers Conference in 2011

In 1971, *Computer Space*, created by Nolan Bushnell and Ted Dabney, was the first commercially sold, coin-operated video game. It used a black-and-white television for its display, and the computer system was made of 74 series TTL chips.[8] The game was featured in the 1973 science fiction film *Soylent Green*. *Computer Space* was followed in 1972 by the Magnavox Odyssey, the first home console. Modeled after a late 1960s prototype console developed by Ralph H. Baer called the "Brown Box", it also used a standard television.[4] [9] These were followed by two versions of Atari's *Pong*; an arcade version in 1972 and a home version in 1975 that dramatically increased video game popularity.[10] The commercial success of *Pong* led numerous other companies to develop *Pong* clones and their own systems, spawning the video game industry.[11]

A flood of *Pong* clones eventually led to the video game crash of 1977, which came to an end with the mainstream success of Taito's 1978 shooter game *Space Invaders*,[12] marking the beginning of the golden age of arcade video games and inspiring dozens of manufacturers to enter the market.[12] [13] The game inspired arcade machines to become prevalent in mainstream locations such as shopping malls, traditional storefronts, restaurants, and convenience stores.[14] The game also became the subject of numerous articles and stories on television and in newspapers and magazines, establishing video gaming as a rapidly growing mainstream hobby.[15] [16] *Space Invaders* was soon licensed for the Atari VCS (later known as Atari 2600), becoming the first "killer app" and quadrupling the console's sales.[17] This helped Atari recover from their earlier losses,[18] and in turn the Atari VCS revived the home video game market during the second generation of consoles, up until the North American video game crash of 1983.[19] The home video game industry was revitalized shortly afterwards by the widespread success of the Nintendo Entertainment System,[20] which marked a shift in the dominance of the video game industry from the United States to Japan during the third generation of consoles.[21]

Overview

Platforms

The term "platform" refers to the specific combination of electronic components or computer hardware which, in conjunction with software, allows a video game to operate.[22] The term "system" is also commonly used.

In common use a "PC game" refers to a form of media that involves a player interacting with a IBM PC compatible personal computer connected to a video monitor. A "console game" is played on a specialized electronic device that connects to a common television set or composite video monitor. A "handheld" gaming device is a self contained electronic device that is portable and can be held in a user's hands. "Arcade game" generally refers to a game played on an even more specialized type of electronic device that is typically designed to play only one game and is encased in a special cabinet. These distinctions are not always clear and there may be games that bridge one or more platforms. In addition to personal computers, there are multiple other devices which have the ability to play games but are not dedicated video game machines, such as mobile phones, PDAs and graphing calculators.

With the advent of social networking and other online applications hubs, the term "platform" started being used to refer to the online service within which the game is played, regardless of the actual hardware on which it is executed. A game's platform could simply be "Facebook", whether it is played on a Windows PC, Mac, Smart TV, or smartphone.

Genres

A video game, like most other forms of media, may be categorized into genres based on many factors such as method of game play, types of goals, art style and more. Because genres are dependent on content for definition, genres have changed and evolved as newer styles of video games have come into existence. Ever advancing technology and production values related to video game development have fostered more life-like and complex games which have in turn introduced or enhanced genre possibilities (e.g., virtual pets), pushed the boundaries of existing video gaming or in some cases add new possibilities in play (such as that seen with titles specifically designed for devices like Sony's EyeToy). Some genres represent combinations of others, such as massively multiplayer online role-playing games, or, more commonly, MMORPGs. It is also common to see higher level genre terms that are collective in nature across all other genres such as with action, music/rhythm or horror-themed video games.

Classifications

Casual games

Casual games derive their name from their ease of accessibility, simple to understand gameplay and quick to grasp rule sets. Additionally, casual games frequently support the ability to jump in and out of play on demand. Casual games as a format existed long before the term was coined and include video games such as Solitaire or Minesweeper which can commonly be found pre-installed with many versions of the Microsoft Windows operating system.

Examples of genres within this category are hidden object, match three, time management, tetris or many of the tower defense style games. Casual games are generally sold through online retailers such as PopCap, Zylom, Vans Video Games and GameHouse or provided for free play through web portals such as Newgrounds or AddictingGames.

While casual games are most commonly played on personal computers, cellphones or PDAs, they can also be found on many of the on-line console system download services (e.g., Xbox Live, PlayStation Network, or WiiWare).

Serious games

Serious games are games that are designed primarily to convey information or a learning experience of some sort to the player. Some serious games may even fail to qualify as a video game in the traditional sense of the term. Educational software does not typically fall under this category (e.g., touch typing tutors, language learning, etc.) and the primary distinction would appear to be based on the title's primary goal as well as target age demographics. As with the other categories, this description is more of a guideline than a rule.

Serious games are games generally made for reasons beyond simple entertainment and as with the core and casual games may include works from any given genre, although some such as exergames, educational games, or propaganda games may have a higher representation in this group due to their subject matter. These games are typically designed to be played by professionals as part of a specific job or for skill set improvement. They can also be created to convey social-political awareness on a specific subject.

One of the longest running serious games franchises would be Microsoft Flight Simulator first published in 1982 under that name. The United States military uses virtual reality based simulations, such as VBS1 for training exercises,[23] as do a growing number of first responder roles (e.g., police, fire fighter, EMT).[24] One example of a non-game environment utilized as a platform for serious game development would be the virtual world of *Second Life*, which is currently used by several United States governmental departments (e.g., NOAA, NASA, JPL), Universities (e.g., Ohio University, MIT) for educational and remote learning programs[25] and businesses (e.g., IBM, Cisco Systems) for meetings and training.[26]

Tactical media in video games plays a crucial role in making a statement or conveying a message on important relevant issues. This form of media allows for a broader audience to be able to receive and gain access to certain information that otherwise may not have reached such people. An example of tactical media in video games would be newsgames. These are short games related to contemporary events designed to illustrate a point.[27] For example, Take Action Games is a game studio collective that was co-founded by Susana Ruiz and has made successful serious games. Some of these games include *Darfur is Dying*, *Finding Zoe*, and *In The Balance*. All of these games bring awareness to important issues and events in an intelligent and well thought out manner.[28]

Educational games

On September 23, 2009, U.S. President Barack Obama launched a campaign called "Educate to Innovate" aimed at improving the technological, mathematical, scientific and engineering abilities of American students. This campaign states that it plans to harness the power of interactive games to help achieve the goal of students excelling in these departments.[29] [30] This campaign has stemmed into many new and exciting opportunities for the video game realm and has contributed to many new competitions. Some of these competitions include the Stem National Video Game Competition and the Imagine Cup.[31] [32] Both of these examples are events that bring a focus to relevant and important current issues that are able to be addressed in the sense of video games to educate and spread knowledge in a new form of media. www.NobelPrize.org uses games to entice the user to learn about information pertaining to the Nobel prize achievements while engaging in a fun to play video game.[33] There are many different types and styles of educational games all the way from counting to spelling to games for kids and games for adults. Some other games do not have any particular targeted audience in mind and intended to simply educate or inform whoever views or plays the game.

Development

Video game development and authorship, much like any other form of entertainment, is frequently a cross-disciplinary field. Video game developers, as employees within this industry are commonly referred, primarily include programmers and graphic designers. Over the years this has expanded to include almost every type of skill that one might see prevalent in the creation of any movie or television program, including sound designers, musicians, and other technicians; as well as skills that are specific to video games, such as the game designer. All of these are managed by producers.

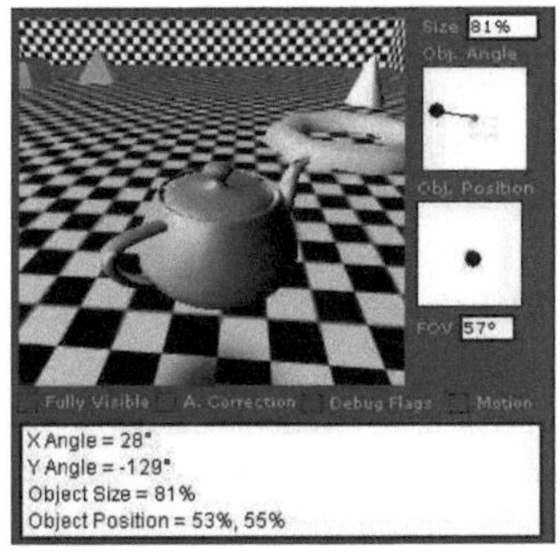

Developers use various tools to create video games, here an editor to fine tune the virtual camera system.

In the early days of the industry, it was more common for a single person to manage all of the roles needed to create a video game. As platforms have become more complex and powerful in the type of material they can present, larger teams have been needed to generate all of the art, programming, cinematography, and more. This is not to say that the age of the "one-man shop" is gone, as this is still sometimes found in the casual gaming and handheld markets,[34] where smaller games are prevalent due to technical limitations such as limited RAM or lack of dedicated 3D graphics rendering capabilities on the target platform (e.g., some cellphones and PDAs).

With the growth of the size of development teams in the industry, the problem of cost has increased. Development studios need to be able to pay their staff a competitive wage in order to attract and retain the best talent, while publishers are constantly looking to keep costs down in order to maintain profitability on their investment. Typically, a video game console development team can range in sizes of anywhere from 5 to 50 people, with some teams exceeding 100. In May 2009, one game project was reported to have a development staff of 450.[35] The growth of team size combined with greater pressure to get completed projects into the market to begin recouping production costs has led to a greater occurrence of missed deadlines and unfinished products.

Downloadable content

A newer phenomenon of withholding content from the game and then releasing it at a later date for additional funds not factored into the retail price began with digital video game distribution known as Downloadable Content (also known colloquially as 'DLC'). Studios may choose to utilize this to issue original content after the game is released, such as Rockstar Games with Grand Theft Auto IV, or Bethesda with Fallout 3, yet often opt instead to create content before the game is released to be intentionally withheld, such as Activision and Treyarch with Call of Duty (Despite knowing well public affection for a 'Zombies' game mode, it was shipped with only one map for this game mode, and periodically releases new ones with $10 DLC).[36] [37] [38]

Modifications

Many games produced for the PC are designed such that technically oriented consumers can modify the game. These mods can add an extra dimension of replayability and interest. Developers such as id Software, Valve Software, Crytek, Bethesda, Epic Games and Blizzard Entertainment ship their games with some of the development tools used to make the game, along with documentation to assist mod developers. The Internet provides an inexpensive medium to promote and distribute mods, and they may be a factor in the commercial success of some games.[39] This allows for the kind of success seen by popular mods such as the *Half-Life* mod *Counter-Strike*.

Cheating

Cheating in computer games may involve cheat codes and hidden spots implemented by the game developers,[40] [41] modification of game code by third parties,[42] [43] or players exploiting a software glitch. Modifications are facilitated by either cheat cartridge hardware or a software trainer.[42] Cheats usually make the game easier by providing an unlimited amount of some resource; for example weapons, health, or ammunition; or perhaps the ability to walk through walls.[41] [42] Other cheats might give access to otherwise unplayable levels or provide unusual or amusing features, like altered game colors or other graphical appearances.

Glitches

Software errors not detected by software testers during development can find their way into released versions of computer and video games. This may happen because the glitch only occurs under unusual circumstances in the game, was deemed too minor to correct, or because the game development was hurried to meet a publication deadline. Glitches can range from minor graphical errors to serious bugs that can delete saved data or cause the game to malfunction. In some cases publishers will release updates (referred to as *patches*) to repair glitches. Sometimes a glitch may be beneficial to the player, these are often referred to as exploits.

Example of a color glitch from the video game *Ultimate Mortal Kombat 3*

Easter eggs

Easter eggs are hidden messages or jokes left in games by developers that are not part of the main game.

Theory

Although departments of computer science have been studying the technical aspects of video games for years, theories that examine games as an artistic medium are a relatively recent development in the humanities. The two most visible schools in this emerging field are ludology and narratology. Narrativists approach video games in the context of what Janet Murray calls "Cyberdrama". That is to say, their major concern is with video games as a storytelling medium, one that arises out of interactive fiction. Murray puts video games in the context of the Holodeck, a fictional piece of technology from *Star Trek*, arguing for the video game as a medium in which we get to become another person, and to act out in another world.[44] This image of video games received early widespread popular support, and forms the basis of films such as *Tron*, *eXistenZ* and *The Last Starfighter*.

Ludologists break sharply and radically from this idea. They argue that a video game is first and foremost a game, which must be understood in terms of its rules, interface, and the concept of play that it deploys. Espen J. Aarseth argues that, although games certainly have plots, characters, and aspects of traditional narratives, these aspects are incidental to gameplay. For example, Aarseth is critical of the widespread attention that narrativists have given to the heroine of the game *Tomb Raider*, saying that "the dimensions of Lara Croft's body, already analyzed to death by film theorists, are irrelevant to me as a player, because a different-looking body would not make me play differently... When I play, I don't even see her body, but see through it and past it."[45] Simply put, ludologists reject traditional theories of art because they claim that the artistic and socially relevant qualities of a video game are primarily determined by the underlying set of rules, demands, and expectations imposed on the player.

While many games rely on emergent principles, video games commonly present simulated story worlds where emergent behavior occurs within the context of the game. The term "emergent narrative" has been used to describe how, in a simulated environment, storyline can be created simply by "what happens to the player."[46] However, emergent behavior is not limited to sophisticated games. In general, any place where event-driven instructions occur for AI in a game, emergent behavior will exist. For instance, take a racing game in which cars are programmed to avoid crashing, and they encounter an obstacle in the track: the cars might then maneuver to avoid the obstacle causing the cars behind them to slow and/or maneuver to accommodate the cars in front of them and the obstacle. The programmer never wrote code to specifically create a traffic jam, yet one now exists in the game.

Social aspects

Demographics

The November 2005 Nielsen Active Gamer Study, taking a survey of 2,000 regular gamers, found that the U.S. games market is diversifying. The age group among male players has expanded significantly in the 25–40 age group. For casual online puzzle-style and simple mobile cell phone games, the gender divide is more or less equal between males and females. Females have also been found to show an attraction to online multi-player games where there is a communal experience. More recently there has been a growing segment of female players engaged with the aggressive style of games historically considered to fall within traditionally male genres (e.g., first-person shooters). According to the ESRB almost 41% of PC gamers are women[47] (see Girl gamer).

When comparing today's industry climate with that of 20 years ago, women and many adults are more inclined to be using products in the industry. While the market for teen and young adult men is still a strong market, it is the other demographics which are posting significant growth. The Entertainment Software Association (ESA) provides the following summary for 2011 based on a study of almost 1,200 American households carried out by Ipsos MediaCT:[48]

- The average gamer is 37 years old and has been playing for 12 years. Eighty-two percent of gamers are 18 years of age or older.
- Forty-two percent of all players are women and women over 18 years of age are one of the industry's fastest growing demographics.
- Today, adult women represent a greater portion of the game-playing population (37 percent) than boys age 17 or younger (13 percent).
- Twenty-nine percent of game players are over the age of 50, an increase from nine percent in 1999.
- Sixty-five percent of gamers play games with other gamers in person.
- Fifty-five percent of gamers play games on their phones or handheld device.

A 2006 academic study, based on a survey answered by 10.000 gamers, identified the gaymers (gamers that identify as LGBT) as a significant demographic group. [49] [50] [51] A followup survey in 2009 studied the purchase habits and content preferences of people in the group.[52] [53] [54]

Based on the study by NPD group in 2011, about 91 percent of kids aged 2–17 play games.[55]

Multiplayer

Video gaming has traditionally been a social experience. Multiplayer video games are those that can be played either competitively, sometimes in Electronic Sports, or cooperatively by using either multiple input devices, or by hotseating. *Tennis for Two*, arguably the first video game, was a two player game, as was its successor *Pong*. The first commercially available game console, the Magnavox Odyssey, had two controller inputs.

Since then, most consoles have been shipped with two or four controller inputs. Some have had the ability to expand to four, eight or as many as 12 inputs with additional adapters, such as the Multitap. Multiplayer arcade games typically feature play for two to four players, sometimes tilting the monitor on its back for a top-down viewing experience allowing players to sit opposite one another.

Many early computer games for non-PC descendant based platforms featured multiplayer support. Personal computer systems from Atari and Commodore both regularly featured at least two game ports. PC-based computer games started with a lower availability of multiplayer options because of technical limitations. PCs typically had either one or no game ports at all. Network games for these early personal computers were generally limited to only text based adventures or MUDs that were played remotely on a dedicated server. This was due both to the slow speed of modems (300-1200-bit/s), and the prohibitive cost involved with putting a computer online in such a way where multiple visitors could make use of it. However, with the advent of widespread local area networking technologies and Internet based online capabilities, the number of players in modern games can be 32 or higher, sometimes featuring integrated text and/or voice chat. MMOs can offer extremely high numbers of simultaneous players; *Eve Online* set a record with 54,446 players on a single server in 2010.[56]

Benefits

It has been shown that action video game players have better hand-eye coordination and visuo-motor skills, such as their resistance to distraction, their sensitivity to information in the peripheral vision and their ability to count briefly presented objects, than nonplayers.[57]

Researchers found that such enhanced abilities could be acquired by training with action games, involving challenges that switch attention between different locations, but not with games requiring concentration on single objects. It has been suggested by a few studies that online/offline video gaming can be used as a therapeutic tool in the treatment of different mental health concerns.

In Steven Johnson's book, *Everything Bad Is Good for You*, he argues that video games in fact demand far more from a player than traditional games like *Monopoly*. To experience the game, the player must first determine the objectives, as well as how to complete them. They must then learn the game controls and how the human-machine interface works, including menus and HUDs. Beyond such skills, which after some time become quite fundamental and are taken for granted by many gamers, video games are based upon the player navigating (and eventually mastering) a highly complex system with many variables. This requires a strong analytical ability, as well as flexibility and adaptability. He argues that the process of learning the boundaries, goals, and controls of a given game is often a highly demanding one that calls on many different areas of cognitive function. Indeed, most games require a great deal of patience and focus from the player, and, contrary to the popular perception that games provide instant gratification, games actually delay gratification far longer than other forms of entertainment such as film or even many books.[58] Some research suggests video games may even increase players' attention capacities.[59]

Learning principles found in video games have been identified as possible techniques with which to reform the U.S. education system.[60] It has been noticed that gamers adopt an attitude while playing that is of such high concentration, they do not realize they are learning, and that if the same attitude could be adopted at school, education would enjoy significant benefits.[61] Students are found to be "learning by doing" while playing video games while fostering creative thinking.[62]

The U.S. Army has deployed machines such as the PackBot and UAV vehicles, which make use of a game-style hand controller to make it more familiar for young people.[63]

According to research discussed at the 2008 Convention of the American Psychological Association, certain types of video games can improve the gamers' dexterity as well as their ability to problem-solve. A study of 33 laparoscopic surgeons found that those who played video games were 27 percent faster at advanced surgical procedures and made 37 percent fewer errors compared to those who did not play video games. A second study of 303 laparoscopic surgeons (82 percent men; 18 percent women) also showed that surgeons who played video games requiring spatial skills and hand dexterity and then performed a drill testing these skills were significantly faster at their first attempt and across all 10 trials than the surgeons who did not play the video games first.[64]

Whilst many studies have detected superior mental aptitudes amongst habitual gamers, research by Walter Boot at the University of Illinois found that non-gamers showed no improvement in memory or multitasking abilities after 20 hours of playing three different games. The researchers suggested that "individuals with superior abilities are more likely to choose video gaming as an activity in the first place".[65] *Unreliable - Discuss*

Controversy

Like related forms of media, computer and video games have been the subject of frequent controversy and censorship, due to the depiction of graphic violence, sexual themes, advergaming (a form of advertising in games), consumption of drugs, consumption of alcohol or tobacco, propaganda, or profanity in some games. Among others, critics of video games often include parents' groups, politicians, organized religious groups, and other advocacy groups. Various games have been accused of causing addiction and even violent behavior,[66] [67] though how much ground this holds is debatable.[68] [69] "Video game censorship" is defined as the use of state or group power to control the playing, distribution, purchase, or sale of video games or computer games. Video game controversy comes in many forms, and censorship is a controversial subject. Proponents and opponents of censorship are often very passionate about their individual views.

Various national content rating organizations, such as the Entertainment Software Ratings Board or ESRB in North America, rate software for certain age groups and with certain content warnings. Some of these organizations are optional industry self-regulation (such as the ESRB), while others are part of national government censorship organizations. Most video games display their rating on the front side of their packaging. However, parents are not always aware of the existence of these ratings. Not all ratings are considered accurate. Organizations such as *What They Play* and *Common Sense Media* aim to provide guidance and advice for parents.[70]

Ratings and censorship

ESRB

The Entertainment Software Rating Board (ESRB) gives videogames maturity ratings based on their content. For example, a game might be rated *T* for *Teen* if the game contained obscene words or violence. If a game contains explicit violence or sexual themes, it is likely to receive an *M* for *Mature* rating, which means that no one under 17 should play it. There is a rated "A/O" games for "Adults Only" these games have massive violence or nudity. There are no laws that prohibit children from purchasing "M" rated games in the United States. Laws attempting to prohibit minors from purchasing "M" rated games were established in California, Illinois, Michigan, Minnesota, and Louisiana, but all were overturned on the grounds that the establishment of such laws were in violation of a child's First Amendment rights. However, many stores have opted to not sell such games to children anyway. Of course, video game laws vary from country to country. One of the most controversial games of all time, *Manhunt 2* by Rockstar Studios, was given an AO rating by the ESRB until Rockstar could make the content more suitable for a mature audience.

Video game manufacturers usually exercise tight control over the games that are made available on their systems, so unusual or special-interest games are more likely to appear as PC games. Free, casual, and browser-based games are usually played on available computers, mobile phones, or PDAs.

PEGI

Pan European Game Information (PEGI) is a system that was developed to standardize the game ratings in all of Europe (not just European Union, although the majority are EU members), the current members are: all EU members, except Germany and the 10 accession states; Norway; Switzerland. Iceland is expected to join soon, as are the 10 EU accession states. For all PEGI members, they use it as their sole system, with the exception of the UK, where if a game contains certain material,[71] it must be rated by BBFC. The PEGI ratings are legally binding in Vienna and it is a criminal offence to sell a game to someone if it is rated above their age.[72]

Germany: BPjM and USK

Stricter game rating laws mean that Germany does not operate within the PEGI. Instead, they adopt their own system of certification which is required by law. The Unterhaltungssoftware Selbstkontrolle (*USK* or Voluntary Certification of Entertainment Software) checks every game before release and assigns an age rating to it - either none (white), 6 years of age (yellow), 12 years of age (green), 16 years of age (blue) or 18 years of age (red). It is forbidden for anyone, retailers, friends or parents alike, to allow a child access to a game for which he or she is underage. If a game is particularly violent, it may be referred to the BPjM (*Bundesprüfstelle für jugendgefährdende Medien* - Federal Verification Office for Child-Endangering Media) who may opt to place it on the *Index* upon which the game may not be sold openly or advertised in the open media. Unofficially, the titles are not "banned" - adult gamers are still technically free to obtain the titles by other means, although it is still considered a felony to supply these titles to a child.

Commercial aspects

Game sales

A typical retail display (in Geneva, Switzerland) with a large selection of games for several major consoles

The three largest producers of and markets for computer and video games (in order) are North America (US and Canada), Japan and the United Kingdom. Other significant markets include Australia, Spain, Germany, South Korea, Mexico, France and Italy.[73] Both India and China are considered emerging markets in the video game industry and sales are expected to rise significantly in the coming years. Irish are the largest per capita consumers of video games.[74]

Sales of different types of games vary widely between these markets due to local preferences. Japanese consumers tend to purchase much more console games than computer games, with a strong preference for games catering to local tastes. Another key difference is that, despite the decline of arcades in the Western world (see Golden age of video arcade games), arcade games remain the largest sector of the Japanese gaming industry. As of 2009, the Japanese gaming industry is worth $20 billion, including $6 billion from arcades, $3.5 billion from console game sales, and $2 billion from mobile game sales.[75] In South Korea, computer games are generally preferred over console games, especially MMORPG games and real-time strategy games. There are over 20,000 Internet cafés in South Korea where computer games can be played for an hourly charge. Computer games are also popular in China, which has a PC gaming industry worth $6 billion, the largest in the world.[76] Arcade games are also still a thriving industry in China, though console games have been banned in the country since the early 2000s.[77]

The NPD Group tracks computer and video game sales in the United States. It reported in 2004 that:

- Console and portable software sales: $6.2 billion, up 8% from 2003[78]
- Console and portable hardware and accessory sales: $3.7 billion, down 35% from 2003[78]
- PC game sales: $1.1 billion, down 15% from 2006[79]

PC games that are digitally distributed either directly or by networks such as Steam are not tracked by the NPD, and Valve does not list sales numbers for games downloaded through their service. Unauthorized distribution is also rampant on the PC.[80]

These figures are sales in dollars, not units, Unit shipments for each category were higher than the dollar sales numbers indicate, because more software and hardware was discounted than in 2003. But with the release of the next-generation consoles in 2006, these numbers increased dramatically. The game and film industries are also becoming increasingly intertwined, with companies like Sony having significant stakes in both. A large number of summer blockbuster films spawn a companion game, often launching at the same time to share the marketing costs.

The global market for console games has seen an average of 6.9 percent compound annual growth rate and is expected to become a $34.7 billion market in 2012.Online game sales are expected to grow at a larger rate of 16.9 percent, escalating from $6.6 billion in 2008 to $14.4 billion by 2012. The largest channel for growth, however, is in mobile gaming with a growth rate of 19 percent; estimated to grow from $5.6 billion in 2008 to $13.5 billion in four years.[81]

Criticism

Video games have the problem of regional lockout. In Australia, while most DVD players are sold region-free to accommodate local consumer rights legislation,[82] video game consoles are still sold fully region protected. Some effort has been made to increase awareness of the issue, specifically to Nintendo of Australia,[83] in the form of a formal report outlining the issues, published by Aaron Rex Davies.[84] The report has gone on to gain a lot of attention in the public media.[85]

Museums

There are many video game museums around the world, for example Museum of Soviet Arcade Machines in Moscow[86] or Computer Game Museum in Berlin.[87]

The Smithsonian American Art Museum is running an exhibition on "The Art of Video Games".[88]

See also

- Outline of video games
- Video game etiquette

Notes

[1] "Television gaming apparatus and method" (http://www.freepatentsonline.com/3659285.html). *United States Patents*. . Retrieved 2008-06-25.

[2] *Patrick Stack. "History of video game consoles" (http://web.archive.org/web/20090226064943/http://time.com/time/covers/1101050523/console_timeline) *Time Magazine website* 2005 /

[3] U.S. Patent 2455992 (http://www.google.com/patents?vid=2455992)

[4] "Welcome to Pong-Story - Introduction" (http://www.pong-story.com/intro.htm). PONG-Story.com. . Retrieved 2007-10-24.

[5] "Welcome to... NIMROD!" (http://www.goodeveca.net/nimrod/). . Retrieved 2007-10-24.

[6] Winter, David. "A.S.Douglas' 1952 Noughts and Crosses game" (http://www.pong-story.com/1952.htm). PONG-Story. . Retrieved 2009-07-03.

[7] Rabin, Steve (2005) [2005-06-14]. *Introduction to Game Development*. Massachusetts: Charles River Media. ISBN 1-58450-377-7.

[8] Marvin Yagoda (2008). "1972 Nutting Associates Computer Space" (http://web.archive.org/web/20081228061939/http://www.marvin3m.com/arcade/cspace.htm). Archived from the original (http://marvin3m.com/arcade/cspace.htm) on 2008-12-28. .

[9] Orlando, Greg (2007-05-15). "Console Portraits: A 40-Year Pictorial History of Gaming" (http://www.wired.com/gaming/gamingreviews/multimedia/2007/05/gallery_game_history). Wired News. . Retrieved 2007-10-24.

[10] "History of Gaming - Interactive Timeline of Game History" (http://www.pbs.org/kcts/videogamerevolution/history/timeline_flash.html). PBS. . Retrieved 2007-10-25.

[11] Miller, Michael (2005-04-01). "A History of Home Video Game Consoles" (http://web.archive.org/web/20071012152258/http://www.informit.com/articles/article.aspx?p=378141). InformIT. Archived from the original (http://www.informit.com/articles/article.aspx?p=378141) on 2007-10-12. . Retrieved 2007-10-25.

[12] Jason Whittaker (2004), *The cyberspace handbook*, Routledge, p. 122, ISBN 0-415-16835-X

[13] Kent, Steven L. (2001). *The Ultimate History of Video Games: From Pong to Pokémon*. Three Rivers Press. p. 500. ISBN 0-7615-3643-4.

[14] Edge Staff (2007-08-13). "The 30 Defining Moments in Gaming" (http://www.next-gen.biz/features/30-defining-moments-gaming). *Edge*. Future plc. . Retrieved 2008-09-18.

[15] "Atari Offers Largest Game Library" (http://www.digitpress.com/library/magazines/electronic_games/electronic_games_winter81.pdf). *Electronic Games* **1** (1): 40–41 [41]. Winter 1981. . Retrieved 1 February 2012.

[16] "Players Guide To Electronic Science Fiction Games" (http://www.archive.org/stream/electronic-games-magazine-1982-03/Electronic_Games_Issue_02_Vol_01_02_1982_Mar#page/n35/mode/1up). *Electronic Games* **1** (2): 35–45 [36]. March 1982. . Retrieved 1 February 2012.

[17] "The Definitive Space Invaders" (http://www.nowgamer.com/features/894697/the_definitive_space_invaders_part_1.html). *Retro Gamer* (Imagine Publishing) (41): 24–33. September 2007. . Retrieved 2011-04-20.

[18] Nick Montfort & Ian Bogost (2009), *Racing the beam: the Atari Video computer system* (http://books.google.co.uk/books?id=DqePfdz_x6gC&pg=PA66), MIT Press, p. 66, ISBN 0-262-01257-X, , retrieved 2011-05-01

[19] Jason Whittaker (2004), *The cyberspace handbook*, Routledge, pp. 122–3, ISBN 0-415-16835-X

[20] Consalvo, Mia (2006). "Console video games and global corporations: Creating a hybrid culture" (http://web.archive.org/web/20080228191914/http://intl-nms.sagepub.com/cgi/reprint/8/1/117.pdf) (PDF). *New Media Society* **8** (1): 117–137.

doi:10.1177/1461444806059921. Archived from the original (http://nms.sagepub.com/cgi/content/abstract/8/1/117) on 2008-02-28. .

[21] Arnie Katz, Bill Kunkel, Joyce Worley (August 1988), "Video Gaming World", *Computer Gaming World*, p. 44, "I'm sure you've noticed that I've made no reference to the Nintendo craze that has repeated the Atari and Mattel Phenomenon of 8 years ago. That's because for American game designers the Nintendo is a non-event: virtually all the work to date has been done in Japan. Only the future will tell if the design process ever crosses the Pacific as efficiently as the container ships and the letters of credit now do."

[22] "platform - Definitions from Dictionary.com" (http://dictionary.reference.com/browse/platform). Dictionary.com. . Retrieved 2007-11-03.

[23] "How Virtual Reality Military Applications Work" (http://science.howstuffworks.com/virtual-military.htm). . Retrieved 2009-11-24.

[24] *Anne Derryberry, "Definition of Serious games (http://www.adobe.com/resources/elearning/pdfs/serious_games_wp.pdf) *Adobe.com* August 2007

[25] "Serious Games in Virtual Worlds: The Future of Enterprise Business Intelligence" (http://www.b-eye-network.com/view/4163). . Retrieved 2009-11-24.

[26] "Using Second Life As A Business-To-Business Tool" (http://www.informationweek.com/blog/main/archives/2007/04/using_second_li_2.html). . Retrieved 2009-11-24.

[27] "Newsgames: Procedural Rhetoric meets Political Cartoons" by Mike Treanor and Michael Mateas (http://www.wingchunsantacruz.com/gamesandart/research/newsgames-DiGRA2009.pdf)

[28] TAKE ACTION games (http://takeactiongames.com/TAG/HOME.html)

[29] Educate to Innovate | The White House (http://www.whitehouse.gov/issues/education/educate-innovate)

[30] President Obama Launches "Educate to Innovate" Campaign for Excellence in Science, Technology, Engineering & Math (Stem) Education | The White House (http://www.whitehouse.gov/the-press-office/president-obama-launches-educate-innovate-campaign-excellence-science-technology-en)

[31] Video Game Competitions to Expand Stem Learning in the United States (http://www.prnewswire.com/news-releases/video-game-competitions-to-expand-stem-learning-in-the-united-states-71656557.html)

[32] Imagine Cup Student Competition - Support FAQ's (http://web.archive.org/web/20080822015626/http://imaginecup.com/support/faq.aspx#Question1.1)

[33] Educational Games (http://nobelprize.org/educational/)

[34] "The Edge of Reason?" (http://www.eurogamer.net/articles/the-edge-of-reason). eurogamer.net. . Retrieved 2009-11-16.

[35] "Assassin's Creed II dev team triples in size" (http://www.nowgamer.com/news/513/assassins-creed-ii-triples-size-of-dev-team), Christopher Reynolds, 18 May 2009, NOW Gamer.

[36] (http://www.giantbomb.com/downloadable-content/92-329/), Giant Bomb.

[37] (http://web.archive.org/web/20110105153819/https://orders.bethsoft.com/dlc/), Bethesda.

[38] (http://www.joystiq.com/2011/04/11/call-of-duty-black-ops-escalation-dlc-xbox-release-date/), Gilbert, Ben, 11 Apr 2011, Joystiq.

[39] Hollywood Reporter interviewing Doug Lombardi, Quote: "Mods absolutely helped us drive huge sales to 'Half-Life,'" (http://web.archive.org/web/20080506004712/http://www.hollywoodreporter.com/hr/search/article_display.jsp?vnu_content_id=1000484956).

[40] Vargas, Jose Antonio (2006-08-28). "In Game World, Cheaters Proudly Prosper" (http://www.washingtonpost.com/wp-dyn/content/article/2006/08/27/AR2006082701059.html). Washington Post. . Retrieved 2007-10-24.

[41] 1UP Staff. "Cracking the Code: The Konami Code" (http://www.1up.com/features/cracking-code-konami-code). 1UP.com. . Retrieved 2007-10-24.

[42] Rybka, Jason. "Video Game Cheats and Codes - What Are Cheat Codes?" (http://vgstrategies.about.com/od/faqglossary/a/WhatAreCheats.htm). About.com. . Retrieved 2007-10-24.

[43] Rybka, Jason. "Why Use Cheats and Codes for Console and PC Games?" (http://vgstrategies.about.com/od/basicgamingtipstricks/a/whyusecheats.htm). About.com. . Retrieved 2007-10-24.

[44] Murray, Janet (1998). *Hamlet on the Holodeck*. MIT Press. ISBN 0-262-63187-3.

[45] Aarseth, Espen J. (2004-05-21). "Genre Trouble" (http://www.electronicbookreview.com/thread/firstperson/vigilant). Electronic Book Review. . Retrieved 2006-06-14.

[46] IGN: GDC 2004: Warren Spector Talks Games Narrative (http://xbox.ign.com/articles/502/502409p1.html)

[47] Guy, Hannah (2007). "Women video gamers: Not just solitaire" (http://www.asiancanadian.net/2007/03/women-video-gamers-not-just-solitaire.html). *PC World* Canada. Archived from the original (http://www.pcworld.ca/news/column/a7fe9b8a0a010408019ac931643ebf2c/pg0.htm) on 2007-03-12. .

[48] ESA. "Game Player Data" (http://www.theesa.com/facts/gameplayer.asp). .

[49] Sliwinski, Alexander (2006-06-10). "First-ever survey of gay videogamers" (http://web.archive.org/web/20060618040019/http://www.washingtonblade.com/thelatest/thelatest.cfm?blog_id=7415). Washington Blade. Archived from the original (http://www.washingtonblade.com/thelatest/thelatest.cfm?blog_id=7415) on 2006-06-18. .

[50] Sliwinski, Alexander (2006-06-08). "Gay video game player survey" (http://web.archive.org/web/20090101100249/http://www.innewsweekly.com/innews/?class_code=Ga&article_code=2071). In Newsweekly. Archived from the original (http://www.innewsweekly.com/innews/?class_code=Ga&article_code=2071) on 2009-01-01. .

[51] Sliwinski, Alexander. "Gay gamer survey results with large hetero inclusion" (http://i.joystiq.com/2007/02/26/gay-gamer-survey-results-with-large-hetero-inclusion/). Joystiq. .

[52] Sliwinski, Alexander (2009-10-08). "New 'gaymer' survey explores sexual identity, interest in games" (http://www.joystiq.com/2009/10/08/new-gaymer-survey-explores-sexual-identity-interest-in-games/#continued). Joystiq.com. .

[53] Fahey, Mike (2009-10-08). "What Do Gay Gamers Want From Their Games?" (http://kotaku.com/5377127/what-do-gay-gamers-want-from-their-games). Kotaku.com. .

[54] Hyman, Jamie (2009-09-03). "Orlando student conducts gay gamer survey" (http://www.watermarkonline.com/w-news/orlando/item/403-orlando-student-conducts-gay-gamer-survey). Orlando Watermark. .

[55] "91% Of Kids Aged 2-17 Playing Video Games, Says Report" (http://techcrunch.com/2011/10/11/91-of-kids-aged-2-17-playing-video-games-says-report/?utm_source=feedburner&utm_medium=feed&utm_campaign=Feed:+Techcrunch+(TechCrunch)). . Retrieved October 12, 2011.

[56] CCP reveals new EVE Online record with 54,446 concurrent users (http://www.strategyinformer.com/news/6545/ccp-reveals-new-eve-online-record-with-54446-concurrent-users)

[57] Green, C. Shawn; Bavelier, Daphne (2003). "Action video game modifies visual selective attention". *Nature* **423** (6939): 534–537. doi:10.1038/nature01647. PMID 12774121. Green & Bavelier.

[58] Brain Candy: The New Yorker (http://www.newyorker.com/archive/2005/05/16/050516crbo_books)

[59] Daphne Bavelier *et al.* (2003). "Action video game modifies visual selective attention" (http://web.archive.org/web/20060329174957/http://www.bcs.rochester.edu/people/daphne/GreenandBavelier.pdf) (PDF). Nature/University of Rochester. Archived from the original (http://www.bcs.rochester.edu/people/daphne/GreenandBavelier.pdf) on March 29, 2006. . Retrieved April 29, 2006.

[60] Gee, James Paul (2003). *What Video Games Have to Teach us About Literacy and Learning*. Palgrave Macmillan. ISBN 1-4039-6169-7.

[61] James Paul Gee *et al.* (2007). "Wired 11.05: View" (http://www.wired.com/wired/archive/11.05/view.html). Codenet, Inc.. . Retrieved December 4, 2007.

[62] Glazer, S. (2006). "Video games". *CQ Researcher* **16**: 960–937. cqresrre2006111000.

[63] "iRobot Receives New Military Orders 14 PackBot Robots" (http://www.spacedaily.com/reports/iRobot_Receives_New_Military_Orders_14_PackBot_Robots_999.html). 2007. . Retrieved July 25, 2007.

[64] Playing Video Games Offers Learning Across Life Span (http://newswise.com/articles/view/543366/) Newswise, Retrieved on August 17, 2008.

[65] "One less excuse to play video games" (http://www.newscientist.com/article/mg20026805.700-video-games-dont-train-your-brain.html). *New Scientist* (Reed Business Information) (2680): 23. November 1, 2008. . Retrieved 2008-10-30.

[66] Jacob Hartley Killed Crying Son For Interrupting Video Game, Cops Say (http://www.huffingtonpost.com/2012/04/09/dad-shakes-baby-jacob-david-hartley_n_1412111.html)

[67] Lakeland man charged with killing infant son after video game session interrupted (http://articles.orlandosentinel.com/2012-04-06/news/os-lakeland-dad-infant-dead-20120406_1_hartley-claims-lakeland-man-infant-son)

[68] Dad charged with killing newborn son after video game session interrupted (http://articles.sun-sentinel.com/2012-04-09/news/sfl-video-game-20120409_1_newborn-son-farmville-game-bad-dad)

[69] Dad killed three-week-old son because he CRIED too much during Xbox marathon (http://www.dailymail.co.uk/news/article-2126585/Jacob-David-Hartley-killed-3-week-old-son-Colton-CRIED-Xbox-marathon.html?ito=feeds-newsxml)

[70] "Grand Theft Auto Makers Sued By LA Attorney For Hidden Porn" (http://www.consolewatcher.com/2006/01/grand-theft-auto-makers-sued-by-la-attorney-for-hidden-porn/). Console Watcher. 2006. . Retrieved October 26, 2006.

[71] "Classification: Guidelines: Introduction" (http://www.bbfc.co.uk/classification/guidelines/introduction/). British Board of Film Classification. . Retrieved 23 November 2011. "The BBFC classifies video works which are released as video recordings under the Video Recordings Act 1984. (The video games covered by the VRA are those whose exemption is forfeited under section 2(2) because they depict human sexual activity, gross violence or other matters of concern.)"

[72] "Pan European Game Information" (http://bupp.at/chancen-amp-risiken/jugendschutz/pegi/) (in German). Bundesstelle für die Positivprädikatisierung von Computer- und Konsolenspielen. . Retrieved 23 November 2011. "Eine Kennzeichung mit der PEGI-Altersempfehlung ist in Wien seit April 2008 verpflichtend: Es dürfen nur Spiele verkauft werden, die eine PEGI-Kennzeichnung tragen. Und es dürfen die Spiele nur an Jugendliche verkauft werden, wenn diese zumindest das Alter gemäß PEGI-Kennzeichung erreicht haben. Bis Ende 2009 gilt in der Übergangsfrist alternativ auch die Kennzeichnung der USK."

[73] Computer And Video Games: A British Phenomena Around the World (http://web.archive.org/web/20080624233600/http://www.elspa.com/assets/files/c/computerandvideogamesabritishphenomenonaroundthewo_174.pdf) (PDF)

[74] Global Entertainment and Media Outlook:2009-2013

[75] Sambe, Yukiharu (NaN undefined NaN). "Japan's Arcade Games and Their Technology" (http://www.springerlink.com/content/tkv5171476213645/). *Lecture Notes in Computer Science*. Entertainment Computing– ICEC 2009 **5709**: 338. doi:10.1007/978-3-642-04052-8_62. . Retrieved 25 January 2012.

[76] Usher, William (2012-03-11). "PC Game Sales Top $18.6 Billion In 2011" (http://www.cinemablend.com/games/PC-Game-Sales-Top-18-6-Billion-2011-40337.html). *Gaming Blend*. Cinema Blend. . Retrieved 17 March 2012.

[77] Jou, Eric (March 19, 2012). "The Wonderful and Seedy World of Chinese Arcades" (http://kotaku.com/5894415/the-wonderful-and-seedy-world-of-chinese-arcades/gallery/1). Kotaku. . Retrieved 9 April 2012.

[78] "U.S. video game industry sales dip in 2004" (http://gameinfowire.com/news.asp?nid=5650). Game Info Wire. January 18, 2005. . Retrieved February 12, 2006.

[79] "Sales & Genre data" (http://web.archive.org/web/20060208194822/http://www.theesa.com/facts/sales_genre_data.php). Entertainment Software Association (ESA). 2004. Archived from the original (http://www.theesa.com/facts/salesandgenre.asp) on 8 February 2006. . Retrieved 12 February 2006.
[80] DiCarlo, Lisa (18 July 2005). "Do Game Publishers Ignore Piracy?" (http://web.archive.org/web/20080103014804/http://www.forbes.com/2005/07/18/videogame-piracy-worldwide-cx_ld_0718piracy.html). Forbes.com. Archived from the original (http://www.forbes.com/2005/07/18/videogame-piracy-worldwide-cx_ld_0718piracy.html) on 2008-01-03. . Retrieved 29 October 2007.
[81] Bond, Paul (2008-06-18). "Video game sales on winning streak, study projects" (http://www.reuters.com/article/2008/06/18/us-videogames-idUSN1840038320080618?pageNumber=1&virtualBrandChannel=0). reuters. . Retrieved 2011-03-12.
[82] "High Court of Australia enforcing region free" (http://www.austlii.edu.au/cgi-bin/sinodisp/au/cases/cth/HCA/2005/58.html)
[83] Nintendo Australia (http://www.nintendo.com.au/)
[84] "Research report into excessive delays in release of Wii software by [[Nintendo (http://homepages.ihug.co.nz/~aaronights/2007-8-11_report_into_software_releases.pdf)] in Australia and New Zealand regions"] by Aaron Rex Davies (PDF)
[85] "Australia to Nintendo: Knock It Off" (http://www.aussie-nintendo.com/?v=news&p=16776)
[86] Museum of Soviet arcade machines (http://www.15kop.ru/en/)
[87] Computerspielemuseum - Berlin (http://www.computerspielemuseum.de/index.php?lg=en)
[88] (http://americanart.si.edu/exhibitions/archive/2012/games/)

References

- Lieu, Tina (August 1997). "Where have all the PC games gone?" (http://web.archive.org/web/19980112155115/http://www.cjmag.co.jp/magazine/issues/1997/aug97/0897pcgames.html). Computing Japan. Archived from the original (http://www.cjmag.co.jp/magazine/issues/1997/aug97/0897pcgames.html) on 1998-01-12.
- Costikyan, Greg (1994). "I Have No Words & I Must Design" (http://www.costik.com/nowords.html).
- Crawford, Chris (1982). "The Art of Computer Game Design" (http://library.vancouver.wsu.edu/art-computer-game-design).
- Salen, Katie; Eric Zimmerman (2005). *The Game Design Reader: A Rules of Play Anthology*. The MIT Press. ISBN 0-262-19536-4.
- Smuts, Aaron (2005). "Are Video Games Art?" (http://www.contempaesthetics.org/newvolume/pages/article.php?articleID=299).
- Blodget, Henry (April 12, 2005). "How to Solve China's Piracy Problem" (http://www.slate.com/articles/arts/go_east_young_man/2005/04/how_to_solve_chinas_piracy_problem.html). Slate.com. Retrieved February 12, 2006.
- Winegarner, Beth (January 28, 2005). "Game sales hit record highs" (http://www.gamespot.com/news/game-sales-hit-record-highs-6117438). Gamespot. Retrieved February 12, 2006.
- John Wills (2002-10-01). "Digital Dinosaurs and Artificial Life: Exploring the Culture of Nature in Computer and Video Games". *Cultural Values (Journal for Cultural Research)* (Routledge) **6** (4): 395–417. doi:10.1080/1362517022000047334.
- Williams, J. P., & Smith, J. H., eds. (2007). The players' realm: studies on the culture of video games and gaming. Jefferson, N.C.: McFarland & Co. ISBN 978-0-7864-2832-8

External links

- Video Games (http://www.dmoz.org/Games/Video_Games/) at the Open Directory Project
- The Virtual Museum of Computing (VMoC) (http://archives.icom.museum/vlmp/computing.html)
- Tom Chatfield writes (http://www.prospectmagazine.co.uk/2008/06/rageagainstthemachines/) on the culture and future of video games for Prospect Magazine
- CBC Digital Archives: The Arcade Age (http://archives.cbc.ca/arts_entertainment/media/topics/3607/)
- An interactive video game history timeline (http://www.mauriciogiraldo.com/vgline/beta/)
- Video Games Through the Ages (http://www.life.com/gallery/42862/video-games-through-the-ages#index/0) - slideshow by *Life magazine*
- Computer games museum in Paris (http://www.museedujeuvideo.com/)

Football Manager

Football Manager is a series of football management simulation games developed by Sports Interactive and published by Sega. The game began its life in 1992 as Championship Manager; however, following the break-up of their partnership with original publishers Eidos Interactive, Sports Interactive lost the naming rights and re-branded the game Football Manager with their new publishers Sega.

Football Manager 2005

On 12 February 2004, after splitting from publishers Eidos Interactive, it was announced that Sports Interactive, developers of the *Championship Manager* game, had retained the rights to the source code but not the rights to the title *Championship Manager*, which were held onto by Eidos (who previously acquired the brand rights from Domark upon their merger in 1995). These developments led to a further announcement that future Sports Interactive football management games would be released under the famous *Football Manager* brand name. Whilst the *Championship Manager* series would go on, Eidos no longer had any source code, or, indeed a developer for *Championship Manager*.

Having been left without a publisher for its football management series, Sports Interactive teamed up with Sega and later, in April 2006, Sports Interactive was acquired completely by the publisher in a continuing trend of consolidation within the games industry.

The first game released under the newly acquired Football Manager brand was *Football Manager 2005*. Commonly known as "FM 2005", it competed directly with *Championship Manager 5* from Eidos-funded Beautiful Game Studios.

Football Manager 2005 included an updated user interface, a refined game engine, updated database and competition rules, pre and post-match information, international player news, cup summary news, a 2D match engine, coach reports on squads, job centre for non-playing positions, mutual contract termination, enhanced player loan options, manager "mind games" and various other features.

Football Manager 2005 was released in the UK on 5 November 2004 - closely followed by releases in many other countries around the world - and it became the 5th fastest selling PC game of all time (according to Eurogamer). The Macintosh version of the game comes on the same dual format disk as the Windows version, so its sales are also included.

Football Manager 2006

Football Manager 2006 for Windows and Mac OS was released in the UK on 21 October 2005 (2 weeks earlier than the originally stated 4 November release). On the same day as the game's release, Sports Interactive also released a patch to fix some bugs discovered during the Beta and Gold stages of development. In its first week of release, it became the second-fastest-selling PC game of all-time in the UK.

Essentially just a season update of FM 2005, it does however, include many small adjustments and improvements to the general gameplay. These adjustments include team-talks, simplified training and in-game help screens. As well as this, the game is updated by its many researchers (unpaid fans of the game augmented by in-house collaboration). The database is usually updated twice in the period of the release of the game. The first comes with the game and the second is usually downloadable in February as a free data update to reflect the changes which take place during the winter opening of the FIFA transfer window. As has been customary with the series a beta demo of the game was released on 12 September 2005. This was later followed on 30 September by a gold demo. This is a cut-down, limited time version of the full game which is sent to the game manufacturers.

With a special download from Sports Interactive, you can play as the fictional football team, Harchester United from Sky One's series *Dream Team*.

Football Manager 2006 - Xbox 360

The Xbox 360 version was released on 13 April 2006 and is the first home console game in the *Football Manager* series. The full 50 playable league systems are included, as well as a 250,000-strong player database (very near to the figure of the PC version), but due to the large save files of the game, the Xbox hard disk drive is required. The game is also region free.

This version also makes use of the Xbox Live functionality, allowing players to create online leagues and cups with up to 16 human-controlled teams using team data they have exported from their offline game. Voice chat is fully supported during online play. It has also been confirmed that SI will release new content through the *Marketplace* system.

Football Manager 2007

On 8 June 2006, Sports Interactive published details on *Football Manager 2007*. "*Packed full with more than 100 new features, Football Manager 2007 is set for release on PC, Mac and Intel Mac, as well as the Xbox 360 video game and entertainment system from Microsoft. Football Manager Handheld on Sony's PlayStation Portable handheld entertainment system (PSP) completes the format line-up, with all platforms set for a Christmas 2006 release.*"

On 24 September 2006 it was announced that *Football Manager 2007* would be released on 20 October 2006.

On 1 October 2006, Sports Interactive released a Gold Demo of FM 2007, available in two versions, vanilla and strawberry. Both versions allow users to play 6 months into a season. The strawberry version contains a larger collection of quickstarts so users can try out more leagues. It also contains more graphics than the vanilla version.

As a result of certain retail outlets ignoring the official release date, *Football Manager 2007* was actually released on 18 October 2006.

New Features in the 2007 version of FM include the ability to include pictures for you as the manager; substantially increased media interaction such as approaches from national newspaper journalists about the upcoming match or asking for comments on a player's performance in recent matches; new varying degrees of criticism or praise for players (happy with form or very pleased with form rather than just one generic good term), a similar feature has been included for such actions as admiring players or attempting to unsettle transfer targets. Improvements have also been made to board request interactions.

The main improvement lauded by Sports Interactive for this version of FM is the improved scouting system. Scouts' and coaches' experience is now shown graphically via bar charts and scouts gain knowledge from areas they have scouted previously. To go with this it is now possible to get feeder/parent club status which allows you to either send your players to lower league teams to gain experience, make it easier to get work permits, merchandise sales in other areas etc. or, as a lower league team, to receive reserve and youth team players on free, season long loans and to generate extra revenue through associated 'fees' and friendly matches. In turn the parent club gains scouting knowledge for the area that the feeder club is in if it is in another country than that club.

A patch has been released to combat some of the main bugs in the game, such as unrealistic scout reports and high amounts of injuries sustained whilst playing on the pitch. The latest build of this patch is 7.0.2 94768.

On 27 July, the Football Manager website was updated with a new Football Manager 2008 section.

Football Manager 2008

On 3 October 2007 Sega Europe Ltd and Sports Interactive announced Football Manager 2008 to be released 19 October. However, the release date was brought forward by a day to 18 October 2007, due to some retailers shipping the game early. There were a number of new features: all features in matches seamlessly follow on from the previous one; a mini radar pitch appears during tactical changes; improved International management; new skin; advisor system; improved notebook system; a revamped captain selection; improved board and fan confidence system; transfer centre to track transfers; improved match engine; the player may change pitch dimensions; award system overhaul; in-game all-time best elevens; a revamped finance system; collective win bonuses; FaceGen system for new-gen players; new calendar.

Football Manager 2009

On 3 September 2008, Sports Interactive released a preview video announcing Football Manager 2009 to be released on the 14th November. The main difference from past versions was the inclusion of a 3D match engine for the first time in the game's history. Other new features included the ability to have female managers and staff, a new press conference system, more in-depth feedback from your assistant manager and a more realistic transfer system. The latest incarnation of Football Manager was also released in DVD format for the first time.

The use of online-activation DRM utilised by Sports Interactive resulted in issues for consumers activating either online, or phone. This was the result of a well orchestrated DDoS attack launched upon activation servers and phone lines, resulting in many users on the Windows platform unable to activate and play.

On 17 December 2008, Sports Interactive announced a deal with Arsenal FC to release an exclusive version of Football Manager 2009, containing all official Arsenal squad player pictures, and an exclusive skin in Arsenal's colours.

Shortly following this announcement, Sports Interactive reinforced their pursuit of online retailers by offering Football Manager 2009 through eSellerate for the Mac OS X platform on December 23. This is the first incarnation of the game to be available to global users through online distribution.

Football Manager 2010

On 12 August 2009, Sports Interactive & Sega Europe Ltd. announced that Football Manager 2010 for PC and Apple Macintosh, and Football Manager Handheld 2010 for PlayStation Portable was to be released on 30 October 2009. Include Competitions:Challenge Cup and Challenge Cup qualification, Caribbean Championship

Football Manager has been hugely successful, managing to gain the number 1 spot in the gaming charts. There have been many improvements made to FM10, including the following:-

- A totally revamped user interface making it simpler and quicker to find the information you want.
- Match Analysis tool showing the actions of each player on the pitch, allowing you to pinpoint any team's strengths and weaknesses.
- Making changes from the touchline by "shouting" instructions from the touchline instantly.
- Included two new national teams, Zanzibar and Tuvalu.

Football Manager 2011

FM2011 was released on 5 November 2010.[2] On 23 July 2010 a number of features as well as the design of the box were leaked before an official announcement could be made.[3]

- Improved training options
- More complex module for searching new players
- Improved graphics and 3D view of the game
- Introduction of football agents to the game
- Playing games in evenings (so called night effect)
- Differentiated weather conditions

Football Manager 2012

Football Manager 2012 was released on 21 October 2011.

- Transfers & Contracts - significant changes to the transfer & contract systems
- Scouting improvements - a new much detailed in-game report has been devised
- 3D Match Improvements - new animations, new crowd system, more stadiums
- Manage Anywhere, Anytime - the ability to add or take away playable nations
- Tone - new system which allows you to specify the way you want to say things
- Intelligent Interface - a new adaptive layout system, new filters & more

Football Manager Handheld

FMH was first released on 13 April 2006 on the PSP. This was Sports Interactive's first ever game for a handheld console.

It was designed to be a separate game and play somewhat differently to the PC/Mac versions of the game due to the different nature of the handheld consoles. The game was designed to be more similar in feel and play to earlier Sports Interactive products - being much faster in nature, while retaining the feeling of being a realistic simulation.

In April 2010 the game became available on the iOS products. The iOS version of the game is very similar in nature to that on the PSP however it contains a wholly new UI because of the touch screen nature of those devices.

In April 2012 the game was also made available on Android devices. The Android version of the game is very similar in nature to the iOS version.

Below is a list of leagues that are playable within FMH 2012.

- **Australia** - A-League

- **Belgium** - Jupiler Pro League, Division II
- **Brazil** - Série A, Série B
- **England** - Premier League, Championship, League One, League Two, Blue Square Premier, Blue Square North, Blue Square South
- **France** - Ligue 1, Ligue 2, National
- **Germany** - Bundesliga, 2. Bundesliga, 3. Liga
- **Netherlands** - Eredivisie, Eerste Divisie
- **Italy** - Serie A, Serie B, Serie C1, Serie C2
- **Portugal** - Liga Sagres, Liga Vitalis
- **Scotland** - Premier League, First Division, Second Division, Third Division
- **Spain** - La Liga, Segunda Divisi6n, Segunda Divisi6n B
- **Wales** - Welsh Premier League, Cymru Alliance, Welsh Football League First Division

Football Manager Live

On 20 April 2007, Sega Europe Ltd and Sports Interactive released details for Football Manager Live which will be a brand new massively multiplayer online game. Released on 4 November 2008, Football Manager Live has its roots set in the Football Manager series, but is designed specifically as a massively multiplayer online game.

On the 12th of December 2008 the first Community Day was arranged by Sports Interactive and Sega. The future of the series and many other matters were discussed with a number of fans who decided to show up for the event.[4]

On the 14th April 2011, it was announced that Football Manager Live would be discontinued after all 'Gameworlds' (Servers) completed their next season (28 day period) - due around the end of May 2011.

Influence

Football Manager has been recognized by real-life football clubs as a source for scouting players. In 2008, Everton FC signed a deal with Sports Interactive allowing them to use the game's database to scout players and opposition.[5]

A video documentary entitled "Football Manager: More Than Just A Game" was produced by journalism graduate and MA student Stephen Milnes and released in October 2010. [6]

Football Manager will celebrate its twentieth birthday this summer (2012), and the people at BackStage Press are already preparing a book with title "Football Manager Stole My Life" to celebrate this remarkable milestone. Part of the book will include stories about how the game has taken over one's life. [7]

See also

- Sports management games

References

[1] FM2009 & FMH2009 (video announcement) (http://www.youtube.com/watch?v=77URmWUc-1k)

[2] Football Manager 2011 release date officially announced (http://community.sigames.com/showthread.php/226268-Football-Manager-2011-release-date)

[3] FM 2011 leaked details on 23 July 2010 (http://www.football-manager.info/newsroom/news/Football-Manager-2011-available-as-of-29th-of-October_.html)

[4] FM Live Community Day in London - Report by CM Revolution (http://www.football-manager.info/articles/read/FM-Live-Community-Day-_-London.html)

[5] Everton signs Football Manager database (http://www.eurogamer.net/articles/everton-signs-football-manager-database)

[6] Football Manager : More Than Just A Game (video documentary) (http://www.fmscout.com/i-1005-Football-Manager-More-Than-Just-A-Game.html)

[7] Football Manager Stole My Life (20th anniversary book) (http://www.fmscout.com/a-football-manager-stole-my-life-20th-anniversary-book.html)

External links

- Official website (http://www.footballmanager.com/)

Championship Manager

Championship Manager	
Developer(s)	Intelek
Publisher(s)	Domark
Designer(s)	Paul Collyer Oliver Collyer
Platform(s)	DOS, Amiga, Atari ST
Release date(s)	September 1, 1992
Genre(s)	Sports
Mode(s)	Single player, hotseat-multiplayer
Media/distribution	Floppy disk

DOS cover art

Championship Manager was the very first game in the popular Championship Manager series of association football management simulation games. The game was originally released on the Amiga[1] and Atari ST in September, 1992 and then ported to the PC/DOS platform soon after. The game was written by Paul and Oliver Collyer (the co-founders of Sports Interactive) in their bedroom.

First impressions

The release of this first version of the game was not an outstanding success, and sales were steady rather than spectacular. Reviews ranged from the encouraging to the dismissive. Even by early 1990s standards, the graphics of *Championship Manager* were primitive and there were many other management games available that were much more visually pleasing, such as *Premier Manager* and *The Manager*. Critics berated the lack of any real graphics, other than coloured text on top of a background image, and the complete lack of any sound effects. Also, what became known as the "match engine" was very basic, consisting of a clock, 3 small meters showing each team's possession and lines of text commentary describing the match action. Rival game, *The Manager*, included a small screen showing TV-style clips of match action. Another drawback was the absence of real player names - each team was populated with players generated at random by the computer.

Gameplay

The game featured four playable English divisions (First through to Fourth; the newly-formed FA Premier League did not appear until CM 93/94). Perhaps most notably, each division contained only 20 teams, whereas in real life they contained 22 or 24 teams at that time. Also included were all of the major domestic cups of the time (including the Anglo-Italian Cup) and the 3 major European trophies (including the now defunct Cup Winners' Cup).

Any teams outside of the four playable divisions and all foreign teams had no player names at all. Instead, players were simply called "No.3" or "No.10" depending on which position they played.

One of the most innovative things about the game was the introduction of "average ratings" for players - after each match the performance of every player was graded from 0-10 and as the season went on the player's average rating would allow the manager to easily see how each player was performing.

Other versions

In 1993, Intelek and Ubi Soft used the Collyers' game code to produce a version for the French market, known as *Guy Roux Manager* (named after the legendary AJ Auxerre manager). It was fully localised for France and included the French first and second divisions as playable leagues, and all text and commentary in French. The *Guy Roux Manager* franchise has since become a very popular and long running franchise in France with games on many platforms.

Also in 1993, the first Norwegian version was released, and the year later, an Italian version. While the Norwegian version was very much similar to the English one, in the Italian version you could substitute up to 4 players in any game and the transfer deadline occurred much earlier in the season. Both the Norwegian and Italian versions featured real named players though, adding to the popularity of the game.

Trivia

- Electronic Arts reportedly turned down the chance to publish *Championship Manager* in 1992.
- In an episode of *Football Focus* in the late 1990s, then-Coventry City manager Gordon Strachan was interviewed with a copy of *Championship Manager* clearly visible on his Highfield Road desk in the background.

See also

- *Championship Manager series*
- *Football Manager*

References

[1] "Amiga reviews: Championship Manager '92" (http://amigareviews.classicgaming.gamespy.com/champi92.htm). .

External links

- Reviews at classicgaming.com (http://www.classicgaming.com/amigareviews/champi92.htm) - A website containing several old magazine reviews of Championship Manager
- Official game info (http://www.sigames.com/softography.php?type=view&id=1) - Championship Manager 1 section on the official Sports Interactive website

Sports_Interactive

Type	Subsidiary
Industry	Interactive entertainment
Founded	1994
Founder(s)	Paul and Oliver Collyer MBEs
Headquarters	Islington, London, United Kingdom
Key people	Miles Jacobson OBE (Studio Director)
Products	Football Manager series Football Manager Live
Employees	+70 [1]
Parent	Sega
Website	www.sigames.com [2]

Sports Interactive Limited is a computer games development company currently based in Old Street, Central London and owned by the Japanese software and video game company Sega. It is the developer of the popular games Football Manager, NHL Eastside Hockey Manager and Championship Manager Quiz, and is the former developer of Championship Manager. Each release of Championship Manager/Football Manager has traditionally been one of the fastest selling PC games in the UK with Championship Manager 4 the fastest of all time.

History

Founded by Paul Collyer and Oliver Collyer, the company is commonly abbreviated to "SI" amongst fans of their games. In 2003, it was announced Sports Interactive would no longer develop games under the brand Championship Manager as this was owned by publisher Eidos Interactive. Instead on 12 February 2004 it was announced that Sports Interactive had signed a publishing deal with Sega and would produce its new football management games under the self-owned brand of Football Manager.

AFC Wimbledon

Sports Interactive have had a relationship with AFC Wimbledon since the club's formation: Championship Manager 01-02 contains a pro-AFC Wimbledon message occasionally visible at start-up (along with a similarly rare but now rather dated "Save Chip!" message), and they are currently AFC Wimbledon's main sponsors, with their logo appearing on the team strip.

In 2004, Sports Interactive announced a deal to help develop Markus Heinsohn's Out of the Park Baseball.[3]

Sega

On April 4, 2006 it was announced that Sega Holdings Europe Ltd, holding company for Sports Interactive's publishers Sega, had acquired Sports Interactive.

On January 1, 2007 Sports Interactive announced that they were beginning work on a new game. The only detail revealed has stated that the game will be football-based, this is Football Manager Handheld 06 and so on

Football Manager Live

On April 20, 2007, Sports Interactive announced the release of a "brand new massively multiplayer online football management game" entitled *Football Manager Live*[4] similar to that of Football Manager 2007, but, as the name suggests, played completely online by up to 1000 people within any "gameworld" of which there will be multiple. The game was originally announced to be released in March 2008 but the game was actually released January 23, 2009.[5] The game uses a database comprising over 330,000 football players and 10 playable leagues. In addition, game players can talk to one another via in-game chats.

Games

- *Championship Manager series*
  - *Championship Manager*
  - *Championship Manager '93*
  - *Championship Manager 2*
  - *Championship Manager 96/97*
  - *Championship Manager 97/98*
  - *Championship Manager 3*
  - *Championship Manager: Season 00/01*
  - *Championship Manager: Season 01/02*
  - *Championship Manager 4*
  - *Championship Manager: Season 03/04*
- *Football Manager*
  - *Football Manager 2005* (4 November 2004 for PC/Mac)
  - *Football Manager 2006* (21 October 2005 for PC/Mac & April for PSP)
  - *Football Manager 2007* (20 October 2006 for PC/Mac & PSP)
  - *Football Manager 2008* (October 2007 for PC/Mac & PSP and March 2008 for Xbox 360)
  - *Football Manager 2009* (14 November 2008 for PC/Mac & PSP)
  - *Football Manager 2010* (30 October 2009 for PC/Mac & PSP and April for iPhone)
  - *Football Manager 2011* (5 November 2010 for PC/Mac, December for PSP & iPhone and April for iPad)
  - *Football Manager 2012* (21 October 2011 for PC/Mac, PSP, iPhone & iPad)
- *Football Manager Live* (23 January 2009)
- *Eastside Hockey Manager*

  - *NHL Eastside Hockey Manager* (2 July 2004)
  - *NHL Eastside Hockey Manager 2005* (27 May 2005)
  - *NHL Eastside Hockey Manager 2005 (North American version)* (27 September 2005)
  - *NHL Eastside Hockey Manager 2007* (22 September 2006)
- *Out of the Park Baseball Manager*
  - *Out of the Park Baseball Manager 2006* (31 May 2006)
  - *Out of the Park Baseball Manager 2007* (23 March 2007)

Accolades

In 2010 co-founders of Sports Interactive Paul Collyer and Oliver Collyer were appointed as Members of the Order of the British Empire for their services to the computer game industry.

In 2011 Studio Director of Sports Interactive Miles Jacobson was appointed as an Officer of the Order of the British Empire again for service to the computer game industry.

Staff

- Miles Jacobson (Studio Director)
- Paul Collyer (Chairman)
- Oliver Collyer (Chairman)
- Grant Appleyard (Senior Producer)
- Marc Vaughan (Head of Handheld Development)
- Craig Hunter (Graphics)
- Chris Rice (Graphics)
- Philip Rolfe (Media)
- Mark Woodger (Research and Localisation)
- Dean Gripton (Head Researcher - England)
- Pete Sottrel (Research)
- Simon Tipple (Research)
- Kevin Turner (Programmer)
- Risto Remes (Programmer)
- Ed Lambert (Programmer)
- Richard Jones (Programmer)
- Neil Brock (Community Administrator)
- Rob Barns-Graham (Community Manager)

References

[1] http://www.sigames.com/studio/
[2] http://www.sigames.com/
[3] Out of the Park Baseball (http://www.ootpdevelopments.com/ootp)
[4] Announcement of game (http://www.footballmanagerlive.com/en/article/89/1526.html)
[5] (http://www.sigames.com/softography.php?type=view&id=28)

External links

- Official website (http://http://www.sigames.com)
- Official Football Manager Website (http://footballmanager.com/)
- Official Forums (http://community.sigames.com)

Front_Office_Football

Front Office Football is a sports management game which places the player at the helm of an NFL football team. Designed by Jim Gindin, as part of his one man company, Solecismic Software, it has had six sequels: Front Office Football 2, Front Office Football 2001, Front Office Football 2002, Front Office Football 4, Front Office Football 2004, and Front Office Football 2007. A college football version titled Front Office Football: The College Years was released in 2001. Having received many favorable reviews over the years, it has been called the most realistic American football simulation for the computer.[1]

Front Office Football 2004 was the first version that allowed for multi-player capability, resulting in the creation of several on-line leagues.

Front Office Football Central (FOFC) is an on-line discussion forum dedicated to discussing Front Office Football game issues, other games in the text-sim community, and other topics.

References

[1] "Front Office Football Review", Gamespot, http://www.gamespot.com/pc/sports/frontofficefootball2004/review.html

External links

- Solecismic Software (http://www.solecismic.com/) The official site of Solecismic Software
- Front Office Football Central (FOFC) (http://www.frontofficefootballcentral.com) A general discussion message board devoted to the series
- *Front Office Football 2001* (http://www.mobygames.com/game/front-office-football-2001) at MobyGames

Front Office Football 2007 Reviews

- Gamenikki (9/10) (http://www.gamenikki.com/g3/reviews/indie/Display.php?id=13)
- Operation Sports (9/10) (http://reviews.operationsports.com/reviews.aspx?game=fof2007pc)
- Football Outsiders (http://www.footballoutsiders.com/2007/01/09/ramblings/reviews/4804/)

Lemonade Stand

Lemonade Stand	
Release date(s)	
Genre(s)	Economics
Rating(s)	

Lemonade Stand is a basic economics game created in 1973 by Bob Jamison of the Minnesota Educational Computing Consortium.[1] Charlie Kellner ported the game to the Apple II platform in February 1979. Throughout the 1980s Apple Computer included Lemonade Stand (along with other software) with the purchase of their systems. A 2006 BusinessWeek article incorrectly claimed that Microsoft's J Allard created the game in 1981.[2]

Gameplay

Like most games created for microcomputers in the 1970s, the gameplay is simple. It simulates a child's lemonade stand, where choices made by the player regarding prices, advertising, etc. will determine the success or failure of the enterprise.

The game owed its success to offering just enough variables to make a complex challenge for users, but still providing a simply-grasped addictive introduction to the offsetting priorities facing a business. The choice of the right prices and quantities on the day of a heat-wave could instill the intense satisfaction unique to a greatly profitable private enterprise.

The player is first given a weather report for the day (sunny, cloudy or hot and dry, each accompanied by a color drawing) and is prompted for three values: The number of glasses of lemonade to make, the number of advertising signs, and the cost of lemonade per glass. The program then gives a report of the earnings for that day. A thunderstorm, sometimes occurring on cloudy days and accompanied by a color animation, will void any profits and cause the player to lose any investment for the day. The game can be played either single-player or with up to 30 players (each player is independent and the sales of one do not affect another). The Apple II version included music, with bars from "Singin' in the Rain" and "Summertime" played at appropriate moments.

See also

- *Lemonade Tycoon*, a similar, more-recent game.
- *Dope Wars*, a game with similar mechanics
- *Lemonade Stand*, a multi-player card game based on the computer game[3]

References

[1] http://www.hotud.org/component/content/article/46-simulation/22650

[2] "The Soul Of A New Microsoft" (http://www.businessweek.com/magazine/content/06_49/b4012001.htm). Businessweek.com. 2006-12-04. . Retrieved 2009-09-24.

[3] Cram, Trevor. "Lemonade Stand--Board Game Geek" (http://boardgamegeek.com/boardgame/104655/lemonade-stand). .

External links

- Short history (http://members.chello.at/theodor.lauppert/games/lemonade.htm), by Theodor Lauppert.
- Codenautics' port of the Apple II game to Windows and Mac OS (http://codenautics.com/lemonade/)
- Home of the Underdogs - Entry: Lemonade Stand (PC Remake) (http://www.the-underdogs.info/game.php?id=2961)
- Lemonade Stand and the Apple revolution (http://www.smh.com.au/news/digital-life/how-tos/lemonade-stand-and-the-apple-revolution/2008/12/23/1229998532318.html)
- Lemonade Stand Business Plan to take lessons learned from game and apply in real life (http://www.familycapers.com/activities/activities-projects/lemonadestand)

Government_simulation_game

A **government simulation** or **political simulation** is a game that attempts to simulate the government and politics of all or part of a nation. These games may include geopolitical situations (involving the formation and execution of foreign policy), the creation of domestic political policies, or the simulation of political campaigns.[1] They differ from the genre of classical wargames due to their discouragement or abstraction of military or action elements.

Background

Games based on geopolitics and elections existed long before the emergence of personal computers and their ability to quickly process large amounts of statistical data. One of the earliest such games was *The Game of Politics*, created by Oswald Lord in 1935[2] which remained in print until 1960. In 1954, the board game *Diplomacy* was created, which differs from other wargames in that it features a "negotiation" phase during which players reach agreements with other players, and then execute military moves simultaneously.[3] National politics has remained a vital area of board gaming, with products such as the 1986 board game *Die Macher* featuring elections in Germany,[4] and *Wreck the Nation* which satirizes the politics of the United States under the Bush administration.[5]

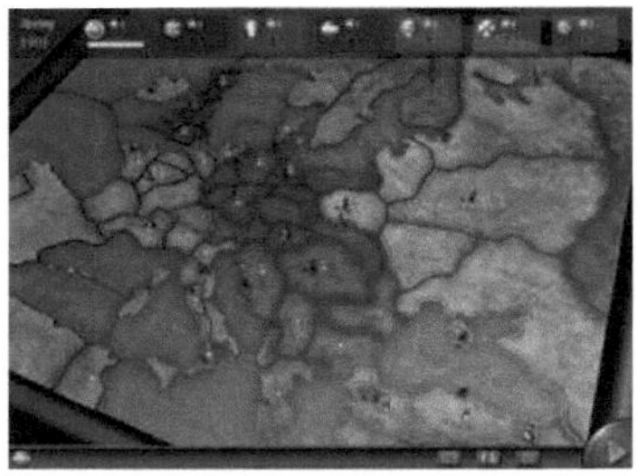

A screenshot from the 2005 computer adaptation of *Diplomacy*, first created as a board game in 1954.

After enjoying years as a play-by-mail game, *Diplomacy* was one of the first games to move to take advantage of e-mail, and continues to be a popular email game as of 2007.[6]

Computer gaming

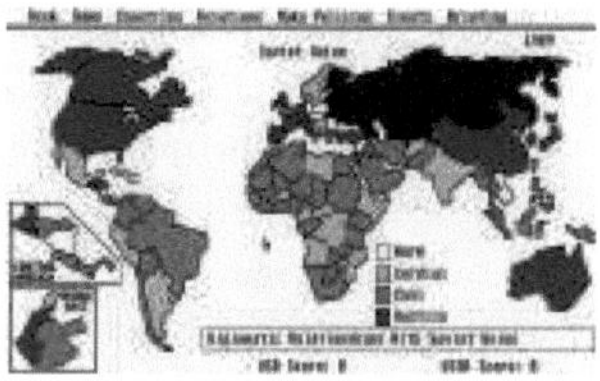

A screenshot from the Atari ST version of *Balance of Power.*

As computers became more sophisticated, games in this genre moved beyond e-mail to more complex simulations. One of the earliest titles in this genre was *Balance of Power*, designed by Chris Crawford and published in 1985. This game features conflict at the height of the Cold War, using political and policy decisions to shape outcomes rather than warfare.[7] [8] In *Balance of Power*, any armed conflict between the player and the opponent superpower results in a nuclear war, which is considered a loss condition.

Other Cold War era games included *Conflict: Middle East Political Simulator* created by Virgin Interactive, Spectrum Holobyte's *Crisis in the Kremlin* and virtually unknown *Hidden Agenda.*

Conflict simulated a hypothetical situation in 1997 in which the player assumed the role of the Israeli Prime Minister and was obligated to employ various diplomatic and covert directives to defeat its rival nations. Surrounded by hostile nations, the player was restrained by a very limited military force and was thereby encouraged to employ peaceful means to remain in power until he acquired more advanced weapons systems and power.[9]

In *Crisis in the Kremlin*, the user could play as the protege of any of the following Soviet politicians: Mikhail Gorbachev of the reformist faction; Yegor Ligachev, leader of the hardline faction; and Boris Yeltsin, who was the prevalent figure of the nationalist faction. The player could use the simulation to test certain strategies to lead the failing Soviet Union into a new era of prosperity or force its dissolution and integration into the new world order. This game introduced the concept of budget management, citizen and faction satisfaction as well as multiple economic values and political spectrum.[10]

In *Hidden Agenda* the user takes the role of the president of Chimerica, a fictional post-revolutionary Central American country, trying to juggle international relations and the needs of the country's citizens.

Early political simulation games were intended more for education than entertainment. In 1987, *On the Campaign Trail* was developed as a tool at Kent State University's political campaign management program, and engaged students in decision-making regarding the campaigns for United States Senate elections between 1970 and 1986.[11] Subsequently, a commercial market developed for packaged games involving elections and campaigns.

A screenshot from Stardock's *Political Machine* game.

The 1992 game *Power Politics* (and, before it, 1981's President Elect)[12] focused on domestic United States political campaigns (but not the running of the country upon election). In 1996, this was adapted to the *Doonesbury Election Game*, designed by Randy Chase (who also did Power Politics) and published by Mindscape, in which players conducted a campaign with the assistance of a pool of advisors selected from characters in the *Doonesbury* comic strip.[13] A successor entitled *Power Politics III* was released in 2005.[14] In 2004, Stardock published *Political Machine*, in which the player steers a candidate through a 41-week election cycle for United States President, developing policies and tailoring talk show appearances and speech content. The game is heavily tied to modern polling methods, using real-time feedback for how campaign strategy impacts polling numbers.[15] In 2006, TheorySpark released *President Forever 2008 + Primaries*, an election simulation game that allows the player to realistically control an entire election campaign through both the Primaries and General Election. *President Forever 2008 + Primaries* itself a follow-up to the highly successful general election sim *President Forever*, released in 2004

Some games in the genre involve enacting policies and budget decisions to sway voters. One such game is *Democracy*, published in 2005 by Positech Games. In *Democracy*, players make decisions during each turn regarding which policies to support. As turns progress, the player views how their favorability rating changes amongst certain types of voters.[16] Candidates make promises before each election, and failure to follow through can result in lower support during the player's reelection campaign.[17] Another is Commander in Chief, produced by Eversim, boasting an array of choices for domestic policy and decisions. Another such game is Tropico.

There can also be found games that puts the player in the seat of a state leader, such as SuperPower, and its sequel, SuperPower 2, whose goals are to produce economic stability and prosperity, but the game mainly revolves around foreign policies, with the abilities to interact with other countries in many ways. The game includes a great number of real-life treaties that influence countries.

Online games

Web-based games such as NationStates, CyberNations and Ars Regendi allow players to manage the day-to-day decisions of individual governments, and compete against rival nations. Less formally structured games are also played out in internet forums, where players manage governments and nations according to a set of agreed rules, such as the MOCpages-based 21st Century World group. Forum-based simulation games have also grown more prominent; one example is *SuperPower: Classic*[18] in which players register, apply for an open nation an avatar (administrator), and then send news article actions which help shape their policies, and aspects such as realism and cooperation are highly promoted.

Other web based games simulate the politics of one specific nation. Players on such games play as fictional politicians and participate in debates, media activity, and simulated elections. One example is *Politics Twist UK*[19] , US Gov Sim [20] and the American Government Simulation (AGS) [21]

Related games

Other construction and management simulations require government management. For example, city-building games such as the *SimCity* series of games published by Maxis simulates the experience of being a mayor. *SimCity* features a real-time environment in which the player can create zones for city development, build roads, power and water utilities, and watch as their city develops based on their decisions. The game was originally published in 1989 and as of 2003 was in its fourth major release.[22]

A mature city from *SimCity 4*.

Strategy games frequently make use of government management challenges. 4X games require the management of a government, be it tribal or interstellar. This includes tasks such as building infrastructure and conducting trade. *Galactic Civilizations II* requires players to manage their approval rating to keep their political party in power. Domestic policy is sometimes abstracted with more emphasis on international conflict. For example, the *Civilization* series gives players total control over resources, and radically restructuring an empire is a matter of clicking a "revolution!" button.

Other strategy games focus on government management to varying degrees. For instance, in the *Hearts of Iron* games (set in World War II) the civilian population is only a factor with partisans and manpower, whereas in *Victoria* a player must not only 'hobnob' and conquer, but implement the Second Industrial Revolution while warding off (ushering in) real political revolutions such as the upheavals of 1848 and communist revolt. Another politically connected game is *Tropico*.

Government and politics have also been incorporated into adventure games. *A Mind Forever Voyaging*, published by Infocom in 1985, was an interactive fiction game in which the player controlled a sentient computer capable of experimenting with potential future scenarios based on varying public policy decisions. Newsweek said of the game, "It isn't '1984,' but in some ways it is even scarier."[23]

Training and education

Beyond entertainment, these games have practical applications in training and education of government personnel. Training simulations have been created for subjects such as managing law enforcement policies (such as racial profiling), the simulation of a military officer's career, and hospital responses to emergency situations.[24]

List of government simulation games

- 4X
- Age of Empires I
- Age of Empires II
- Age of Empires III
- Ars Regendi
- Balance of Power
- Bandit Kings of Ancient China
- Civilization
- Civilization II
- Civilization III
- Civilization IV
- Civilization V
- Civilization Revolution
- Commander in Chief (alternate title: Geo-Political Simulator)[25]
- Conflict: Middle East Political Simulator
- Crisis in the Kremlin
- Crusader Kings II
- Cyber Nations
- CyberJudas
- Destiny of an Emperor
- Democracy
- Europa Universalis III
- Floor 13
- Galactic Civilizations II
- Galava
- Hearts of Iron
- Hearts of Iron II
- Hearts of Iron III
- Hidden Agenda
- Jennifer Government: NationStates
- Nobunaga's Ambition
- Romance of the Three Kingdoms
- Pixel Nations
- Power Politics
- Political Tycoon[26]
- President Elect
- President Forever 2008 + Primaries
- Project Terra
- Republic: The Revolution[27]
- Rulers of Nations
- Sengoku
- Shadow President
- SuperPower
- SuperPower 2
- Supreme Ruler 2010
- Supreme Ruler 2020
- Supreme Ruler Cold War
- Tropico
- Tropico 2
- Tropico 3
- Tropico 4
- The Political Machine
- The Political Machine 2008
- Victoria
- Victoria II

References

[1] Tom Leupold (2004-08-12). "Spot On: Games get political" (http://www.gamespot.com/news/2004/08/12/news_6104371.html). Gamespot. . Retrieved 2007-01-07.

[2] Time Magazine, February 3, 1936, "Monopoly & Politics" (http://www.time.com/time/magazine/article/0,9171,755767-1,00.html)

[3] Allan B. Calhamer, Europa Express #10, "The Roots of Diplomacy" (http://www.diplom.org/Zine/S1998R/Calhamer/invention.html)

[4] Erik Arneson, "Playing Politics" (http://boardgames.about.com/library/weekly/aa061399.htm)

[5] "BuzzFlash Reviews" (http://www.buzzflash.com/store/reviews/274). Buzzflash.com. . Retrieved 2011-01-17.

[6] Jim Burgess, "Play-by-Mail Diplomacy vs Play-by-Email Diplomacy" (http://www.diplomacy-archive.com/resources/postal/pbm-vs-pbem.htm)

[7] Chris Crawford (2003), Chris Crawford on Game Design, ISBN 0-13-146099-4

[8] Robert Mandel, The Journal of Conflict Resolution, Vol. 31, No. 2 (Jun., 1987), "An Evaluation of the 'Balance of Power' Simulation" (http://links.jstor.org/sici?sici=0022-0027(198706)31:2<333:AEOT"O>2.0.CO;2-5), pp. 333-345,

[9] Zzap! Issue 70, February 1991, p.48, "Conflict: the Middle East Political Simulator" (http://web.archive.org/web/20071017063735/http://amigareviews.classicgaming.gamespy.com/conflict.htm)

[10] Social Science Computer Review, Vol. 12, No. 3, 447-448 (1994), "Software Reviews: Crisis in the Kremlin"

[11] Nadine S. Koch, "Winning Is Not the Only Thing 'On the Campaign Trail': An Evaluation of a Micro-Computer Campaign Simulation" (http://links.jstor.org/sici?sici=1049-0965(199112)24:4<694:WINTOT>2.0.CO;2-4&size=LARGE) PS: Political Science and Politics,

Vol. 24, No. 4 (Dec., 1991), pp. 694-698,
[12] "President Elect." (http://www.mobygames.com/game/president-elect) Moby Games (retrieved on January 25th, 2009).
[13] "IGN: The Doonesbury Election Game" (http://pc.ign.com/objects/659/659635.html). Pc.ign.com. 1995-12-30. . Retrieved 2011-01-17.
[14] "Power Politics III (PC)" (http://uk.pc.gamespy.com/pc/power-politics-iii/). GameSpy. . Retrieved 2007-12-22.
[15] Jason Silverman (2004-08-19). "Campaign Game Mimics Real Life" (http://www.wired.com/gaming/gamingreviews/news/2004/08/64625). Wired News. . Retrieved 2007-12-22.
[16] Jess Nickelsen. "Democracy (PC)". NZGamer.com.
[17] "Positech Democracy" (http://www.positech.co.uk/democracy/democracy1.html). . Retrieved 2007-12-06.
[18] Official website of *SuperPower: Classic* (http://itake.se/spcoldwar/index.php?act=idx)
[19] Official website of *Politics Twist UK* (http://politicstwistuk.com/)
[20] Official website of *US Gov Sim* (http://www.govsim.net//)
[21] Official website of *American Government Simulation* (http://www.americangovernmentsimulation.com/forum///)
[22] Tal Blevins (2003-01-14). "Sim City 4 Review" (http://pc.ign.com/articles/382/382583p1.html). IGN. . Retrieved 2007-12-22.
[23] "Ad-Blurbs for A Mind Forever Voyaging" (http://www.mobygames.com/game/a-mind-forever-voyaging/adblurbs). MobyGames. . Retrieved 2007-12-22.
[24] Dave Carey (2007-01-06). "Simulation games help prepare government, unite local businesses" (http://washingtonexaminer.com/simulation-games-help-prepare-government-unite-local-businesses). The Washington Examiner. . Retrieved 2011-03-18.
[25] "Geo-Political Simulator - PC review at Thunderbolt Games" (http://www.thunderboltgames.com/reviews/article/geopolitical-simulator-review-for-pc.html). Thunderboltgames.com. 2008-11-16. . Retrieved 2011-01-17.
[26] "Political Tycoon - PC Review at IGN" (http://pc.ign.com/articles/166/166873p1.html). Pc.ign.com. 2001-08-22. . Retrieved 2011-01-17.
[27] "Republic: The Revolution - Review" (http://play.tm/review/2320/republic-the-revolution/). play.tm. 2003-10-13. . Retrieved 2011-01-17.

External links

- Federal Consortium of Virtual Worlds (http://www.ndu.edu/iCollege/fcvw/index.htm) (National Defense University, Washington, DC)

Article Sources and Contributors

Text_sim *Source*: http://en.wikipedia.org/w/index.php?title=Text_sim *Contributors*: Barkeep49, Brokenyard, Comrade Tux, Cryptic, D.brodale, Deckiller, Excirial, Footypedia, Ian Pitchford, Jam000qaz, Joeandspiro, Maizpan, MrOllie, Notque, Pascal.Tesson, ShelfSkewed, SirFozzie, Swaggs, TheParanoidOne, TizianoF, Warpflyght, 19 anonymous edits

Computer *Source*: http://en.wikipedia.org/w/index.php?title=Computer *Contributors*: 1297, 144.92.164.xxx, 193.203.83.xxx, 24fan24, 7265, 876wer, A D Monroe III, A Softer Answer, A d777, A2raya07, ABShipper, AThing, AXRL, Aaron Schulz, AaronTownsend, Abc753159, Abelson, Abhingeorgegodwin, Abiyoyo, Abner Doon, Academic Challenger, Acasson, Accurizer, Aceofskies05, Acroterion, AdAdAdAd, Adam1213, AdamM, Adashiel, Adnandeura, Adolphus79, Adrian Robson, Adrian.benko, Afghangangster, Ahoerstemeier, Aim Here, Akamad, Akrancis, AlMac, Alan Liefting, Alatius, Aldie, Ale jrb, AlefZet, Alegoo92, Alexmyboy, AlistairMcMillan, Allen Jesus, Alphax, Alwolff55, Amazon10x, AmyzzXX, Anacon, Ancheta Wis, Andoni, Andonic, Andre Engels, Andrew73, Andrewbadr, Andrewpmk, Android79, Andy Janata, Andy24, Andycjp, AngelOfSadness, Angela, Anger22, Angus Lepper, Anmol9999, Anonymous editor, Anshuman.jrt, Antandrus, Antony the genius, Apol0gies, AquaRichy, Arbero, Archer3, ArglebargleIV, Arjun01, Arpingstone, ArsenalTechKB, Arthur Rubin, Arwel Parry, AshLin, Atomaton, Atomice, AtticusX, Aude, Auric, Awien, AxelBoldt, AzaToth, Aztek2313, B. van der Wee, B4hand, BDerrly, BW, Babij, Banes, Barefootguru, BaronLarf, Baseball007, Bastien Sens-Méyé, BazookaJoe, Bcasterline, Bcnfal@hotmail.com, Ben Standeven, Ben-Zin, Ben414, Benched3, Benhoyt, Benny476, Beweird123456, Bhebhe19, BigCow, BigFatBuddha, Bill37212, Birdhombre, Bissinger, Bjarki S, Bjmurph, Bkell, Bkkbrad, Bkonrad, Blackmail, Blacksmith, Blainster, Blaxthos, Blazzer44, Blitz1941, Blueforce4116, Bluemoose, BobShair, Bobblewik, Bobo192, Bogdangiusca, Bonadea, Bongwarrior, Bookandcoffee, Bookinvestor, Bookofjude, BorgQueen, Branddobbe, Brandenads, Branrile09, Breno, Brian0918, BriandaBrain1447, Brianjd, BrokenSegue, Brokenfrog, BrotherFlounder, Brunnock, Brusegadi, Bryan Derksen, Bubba73, Bucketsofg, Buidinhthiem, Butros, C.Fred, C1932, C777, CBDroege, CLW, COMPATT, COMPFUNK2, CSI Laredo, CSWarren, CTanguy, CTho, Cactus.man, Cadiomals, Caknuck, Cal T, Camarcus, Cameron168, Can't sleep, clown will eat me, Canderson7, Cangate, CanisRufus, CapitalR, CaptainVindaloo, Casper2k3, Catdude, CatherineMunro, Causa sui, Cause of death, Cbrodersen, Cbrown1023, Cburnett, Cdc, Cedars, Cellmaker, Cenarium, Cfailde, Cfilorvy, Chameleon, Chappie2006, CharlesGillingham, Cheesewheel, Cheezyp18, Chinneeb, Chowbok, Chris 73, Chris j wood, Chriscm, Chrisjj, Chrislk02, Christian List, Christian Storm, Christopheee, Christopher Parham, Christy747, Chshoaib, Chun-hian, Cicicicico, Cimbalom, Claygate, Cleared as filed, Clockwork Soul, ClockworkSoul, Cma, Cmputer, CoMaDaReInCaRnAtE, CodeCat, Codegrinder, Colin99, Collard, Color probe, Compaqevo, Computerjoe, Conversion script, Cool200, Coolbho3000, Coolcaesar, Coolman435, Crazycomputers, Cremepuff222, Crispichikin, Crusadeonilliteracy, Csarvey, Ctjf83, Curious DGM, Curps, Cursit, Curtis.Everingham, Cutter20, Cverlo, Cy0x, Cyan, Cybercobra, Cybiko123, Cyfal, CyrilB, DJ Clayworth, DRTllbrg, DV8 2XL, DVD R W, DaiTengu, Damicatz, Dan D. Ric, Dan Hickman, Danakil, DangApricot, Dangerousnerd, Daniel C, Daniel Lawrence, DanielCD, Danigoldman, DarkFalls, Darrendeng, DarthVader, Dashes, Dasunt, Daven200520, Daveydweeb, David R. Ingham, Davidgoldner18, Dawn Bard, Dbunds, Dcljr, Dcooper, Delahe15, Delldot, Demmy, DerHexer, Deskana, Dharmabum420, Dhong55, Dhp1080, Diderot, Digitalme, DirkvdM, Disavian, Discospinster, Djembayz, Djhbrown, Dlauri, Dmcq, Dmharvey, Dmn, Dmsar, Doc glasgow, DocWatson42, Dome89, Dominus, Don't fear the reaper, Donald Albury, Donarreiskoffer, Dori, Dr.Bhatta, Drdestiny77, Drummer1508, Duckman89, Duomillia, Dust Filter, Dyl, Dysprosia, ESkog, EagleOne, Eaglesfan2593, Easel3, Eccentrix inc, Ed g2s, Edam, Edcolins, Edgar181, Edits, Edlin, Eiler7, El C, ElTyrant, Elroch, Emersoni, Emoboy 99, Emote, Emre D., Emurph, Epicstonemason, Equendil, Eran of Arcadia, ErrantX, Error5001, Escape Orbit, Eurobas, Evan Robidoux, Evercat, Everyking, Evil Monkey, Evil saltine, Expensivehat, Extra999, Ezrdr, F, FF2010, FT2, Fallout boy, Fantasy, Faucon7, Favonian, Fazdadaz, Feezo, Felixdakat, Ferkelparade, Firsfron, FisherQueen, Flammingo, Flowerparty, Flyguy649, FocalPoint, Foobar, FootholdTechnology, Formatinitials, Fourthords, Foxtrotalpha, FrYGuY, Franamax, Francs2000, Frap, Frazzydee, Freakofnurture, Freakydance, Fredrik, FreplySpang, Frietjes, Frymaster, Fsiler, Funandtrvl, Funnybunny, Furrykef, Fuzheado, GTBacchus, Gaga654, Gaius Cornelius, Galactor213, Gap9551, Gardar Rurak, Garrison Savannah, Gary D Robson, Gary King, Gazpacho, Geeoharee, General Rommel, Geni, Georgy90, Geosultan4, Getsuga, Ghettoblaster, Giftlite, Gilliam, Gimboid13, Gimmetrow, GliderMaven, Globalsolidarity, Glome83, Gmcole, God101, Goel madhur, Gogo Dodo, Golbez, Goldom, Gracenotes, GraemeL, Graham87, Grand Edgemaster, GregAsche, Greswik, Grm wnr, Gscshoyru, Guanaco, Guppy, Gurch, Guy Harris, Guy M, Guy Peters, Gwern, Gwernol, Gwizard, Gyrofrog, Gzornenplatz, HJKeats, Hadal, Haham hanuka, Hallenrm, Hamiltondaniel, Hannes Hirzel, Happyisenough, HarisM, Harmil, Hazmo1, Hdante, Hdt83, HeikoEvermann, Helena srilowa, Helix84, Helixblue, HellRaiserDP, Hemanshu, Henry Flower, Hi-lariousdude22, Hi332211, HiLo48, Highdefinition, Hintha, Hipporoo, Homeless5, Homerjay, Hotshot977, Htaccess, Hu, Hughey, HugoLoris, HunterX, Hurricane111, Hyperboreer88, I luv jonas brothers, IAMARTHUR, Ian slater, Ian13, IanLewis, Idleguy, Ike9898, Ikiroid, Illyria05, Imroy, InShaneee, Indium, Infinity Wasted, Intgr, Invader chris, Invisigoth, Ipigott, Irpatos, Isam, Isotope23, Itscrazyluv, Ixfd64, Izcool, J-Wiki, J.delanoy, J1459, JC Chu, JDspeeder1, JSimmonz, JTN, JYolkowski, Jacek Kendysz, Jackehammond, Jacky man Toronto, Jagged 85, Jaimo01, JamesTeterenko, Jameshater, Jamesk111, Janekm, Janke, Janus303, Jaranda, Jarble, Jarle fagerheim, Jasper Deng, Jaxl, Jbitkill, Jcbutler, JeLuF, JederCoulious, Jeff G., Jeff3000, Jenn0123, Jessesaurus, Jhballard, Jiang, Jim Horning, JimVC3, Jimbreed, JimmyShelter, Jjshapiro, Jkl, Jni, JoanneB, Joe Beaudoin Jr., Joe07734, Joezamboni, Johann Wolfgang, Johantheghost, John, Johnleemk, Johnnyw, Johnrpenner, Jojit fb, Jon Awbrey, JonHarder, Jonathanfspencer, Jorophose, Jose77, Josephs1, Josh Parris, JoshEdgar, JoshuaZ, Joshw101, Jossi, Joyous!, Jpbowen, Jpgordon, Jpisokas, Jpolster2005, Jrauser, Jredmond, Jrockley, Jrpibb, Jtatum, Jtkiefer, Juansempere, Jumbuck, Jus930710, JustPhil, Justin Hirsh, Justin Stafford, Jwissick, Karam.Anthony.K, Karch, Karl2620, Kaseyjean, Katavothron, Katimawan2005, Kazmimi, Kbdank71, KeKe, Keegan, Kelly Martin, KellyCoinGuy, Ken428, KennedyBaird, Kenny sh, KerryO77, Ketiltrout, Kevin B12, Kevin Langendyk, Khin007, Kim Bruning, King of Hearts, KingGrue, Kizor, Kkhairunnisa, Klemen Kocjancic, Klenje, Klingoncowboy4, KnowledgeOfSelf, Koman90, Konstable, Kornxi, Kortsleting, Koyaanis Qatsi, Kozuch, Krawi, Krich, Kubanczyk, Kubigula, KumfyKittyKlub, Kungfuadam, Kurt m 4, KurtRaschke, Kuru, Kwertii, Kx1186, L337p4wn, Lachiester, Lajm, Lappado, Laptop65, Laurinavicius, Leahcim512, Lectonar, Leeyhe, Leigh, Lethe, Leuliett, Levin, Lexi Marie, Lexor, Li-sung, Liftarn, Lightdarkness, Lightmouse, Lights, Ligulem, Lincher, Link5547, Linkspamremover, Linuxbeak, Linuxerist, Litefantastic, LizGere, Llamadog903, Lloydpick, Locos epraix, Looloopoo9, Lord Muck, Luna Santin, Lupin, Lupo, Lysander89, MER-C, MIT Trekkie, MJGR, MONGO, MPerel, MZMcBride, Mac, Madchester, Madman91, Maelor, Magicker71, Magister Mathematicae, Mahagna, Mailer diablo, Majorclanger, Makewater, Maladziec, Malleus Fatuorum, Malo, Manitu, Manjithkaini, MansonP, Manufracture, Maralia, Marcika, Marcus Qwertyus, Marioromeroaguirre, MarkS, Marnanel, Marsheo, Martin g2, MartinDK, Marxmax, Marysunshine, Masiano, Master Jay, Mathboy155, Matheustex, Matilda, Matt Britt, Matt Crypto, MattGiuca, Matthew Fennell, Mattman2593, Maurice Carbonaro, Maustrauser, Mav, MaxD, Maxim, Maximus4140, Maxlaker, MayaSimFan, Mbarbier, McNeight, Mcelite, MeBee, Melaen, Memenen, Mensch, Merovingian, Methnor, Metrax, Mets501, Michael Hardy, Michael Jones jnr, Michael K. Edwards, Michael Zimmermann, Michael93555, Michaelas10, Mightyman67, Mike Rosoft, Mike in Aus, MikeSy, Mikeblas, Mini-Geek, Mirelespm, Misza13, Mjpieters, Mkr10001, Moppet65535, Mortense, Mpradeep, Mr x2, Mr. Billion, Mr. Lefty, Mr. Maupin, MrBoo, MrFish, Mrtomaas, Mudux01, Muffin34, Mukulnryn, Murray Langton, Mushroom, Mxn, NHRHS2010, NSLE, Nachoman-au, Nafsadh, Nageh, Nagytibi, Najoj, Nakon, Nameneko, Nancy, Nanosilver, Nanshu, Nascar1996, Ndp2005goh, Necrowarrio0, Neelix, Neil12, Nelson50, Nervi modest, Netoholic, NewEnglandYankee, Newnew123, Niall2, Nick, NickDanger42, NigelR, Nigelj, Nihiltres, Nike8, Nikola Smolenski, Nilmerg, Ninjagecko, Niteowlneils, No Guru, NoSeptember, Node ue, Noldoaran, Nono64, Norm, NrDg, Nsmith 84, Nucleusboy, Nurg, Nyttend, OOJaxxOo, Obeso24, Oda Mari, Off!, Ohconfucius, Ohnoitsjamie, Ohokohok, Oleg Alexandrov, Oli Filth, OliD, Olorin28, Omicronpersei8, Orderud, OregonD00d, Orioane, Osric, Otets, OtherPerson, OverlordQ, OwenX, Owned3, Oxymoron83, P09ol,, PFHLai, PJM, PaePae, Pagingmrherman, Pagrashtak, Panser Born, Paolo.dL, Papadopa, Papppfaffe, Patrick, Paul August, Pax:Vobiscum, Pegasus1138, Perl87, Peruvianllama, Peterdjones, Petre Buzdugan, Pgk, Phaedriel, Phalacee, Phgao, Philip Trueman, PhilipO, Pholy, Physis, Piccor, Picus viridis, PierreAbbat, Pigsonthewing, Piotrus, Plastikspork, Pmaguire, Pmjjj, Pnm, Poindexter Propellerhead, Poli, Poor Yorick, Pooresd, Pope16, Postdlf, Powo, Pradkart, Prodego, Psy guy, Public Menace, PuzzletChung, Pwner2, Python eggs, Qirex, Qst, Quackor, Quadell, Quarma, Quentin mcalmott, Quiksilviana, Quispiam, Qwerasd1, Qxz, R. S. Shaw, RMuffin, RN1970, RTC, RW Marloe, RaCha'ar, Rabidbuzz, Rac7hel, Radagast83, RadioKirk, Radius, Ragib, RandomP, Rangoon11, RapidR, Rasmus Faber, Raul654, Raven4x4x, Rd232, Rdsmith4, RedWolf, Redthoreau, Reedy, Reisio, Reject, Revolución, RexNL, Rfl, Rgreenday1011, Rhynchosaur, Riana, Rich Farmbrough, Richard001, Richdude24, Rick Sidwell, Rico402, Rieger, Rigadoun, Rilak, Rishi225, Rjsc, RkOrton, Rlinfinity, Rnt20, Robert Brockway, Robert K S, Robert Merkel, RobertG, Robertvan1, Rockhall, Rocky34, Rodrigo braz, RogueMountie, Romanm, Rory096, Roscoe x, Rovibroni, Rowan Moore, RoyBoy, Royote, Rsduhamel, Rubicon, Rudjek, Ruhrjung, Rwwww, RxS, Ryuch, Ryulong, S14jduma, SCEhardt, SCJohnson77, SG, SJP, ST47, Salsb, Saltiem, Sam Hocevar, Sam Li, Sam130132, Samuel, Sander123, Sango123, Saudade7, Saulsinaloa, Saurabh jain999, Sc147, Sceptre, SchreyP, Schzmo, Science History, Scott McNay, Scott Paeth, Sdornan, Sean Whitton, SebastianHelm, Secretlondon, Seewolf, Seibei, Selesti, Seraphim, Serlin, Shadikka, Shadow Android, Shadow1, Shalom Yechiel, Shanel, Shanes, ShaunES, Shauom, Shevonsilva, Shoshonna, Sickbrah, Siddhant, Sietse Snel, Sigma 7, Silentx, Simoes, SimonP, Singsmasta, Singularity, Sir Nicholas de Mimsy-Porpington, SirVulture, Siroxo, SivaKumar, Sjakkalle, Skidude9950, Skraz, Slakr, Slark, Sligocki, Slohar, Smack, Smart Nomad, Smash, Smilliga, Smokizzy, Snozzer, Socalaaron, Soir, SomeStranger, Someone else, Sonny1day, Spaceboy492, Spangineer, Spartan-James, Speaker4000, Spearhead, Specs112, Spelling Corrector, Spencer195, Spiritg1rl95, Splash, Spliffy, Sportzplyr9090, SpuriousQ, Squishy, Srice13, Srikeit, Ssd175, Ssilvers, Steelergolf11, Stemonitis, Stephen Compall, Stephenb, SteveBaker, Stevenj, Stevo1000, Stijn Vermeeren, Stormie, Stormscape, Strait, StuffOfInterest, Subodhdamle, SunCreator, SuperDude115, SuperHamster, Supertouch, Swatjester, Symane, Syvanen, T-rex, T0ny, TERdON, THB, TJDay, TXAggie, Taajikhan, Tabby, Tangotango, Tapir Terrific, Targaryen, Tarret, Tasc, Tastemyhouse, Tawker, Taxman, Tdvance, TeaDrinker, Technion, TedColes, TedE, Tedzdog, Terence, Test2008, Test2010, Tetsuo, Tevildo, TexasAndroid, Thadius856, The Rambling Man, The Stoneman, The informator, The rekcaH, The silent assasin, TheGWO, TheGeneralUser, TheYmode, Thepielord, Theresa knott, Theroadislong, This user has left wikipedia, Thomas Larsen, ThrustVectoring, Thue, Thunderbrand, Thw1309, TigerShark, Tigerhawkvok, Tillmo, Tim1988, TimTIm, Timhowardriley, Timir Saxa, TimmyTimson, Timwi, Titoxd, Tobby72, Tobias Bergemann, Tom harrison, Tom5760, TomTheHand, Tomgally, Tomi T Ahonen, TonyClarke, Torc2, Tosayit, ToxicPlatypus, Tpbradbury, Traroth, TravelinSista, Tregoweth, Trenchcoat99, Trinitymix, Trobert, Trodaikid1983, Trovatore, Trusilver, Truthflux, Ttz642, Tulip19, Tír na nÓg 1982, U, U.Steele, Uartseieu, Ugen64, Ugur Basak, UkPaolo, UncleDouggie, Urod, Useight, Utcursch, Vald, Vanished User 0001, Vanished user 03, Vary, Velella, Vesailius, Viriditas, Virtual Traveler, Vovkav, Vranak, Vulcanstar6, WAS 4.250, Waggers, Wanderingcat, Wanderingstan, Wavelength, Wayfarer, Wayiran, Wayward, Wbm1058, Wernher, West Brom 4ever, Whaa?, Where, WhiteDragon, Whosasking, Widefox, Wiki alf, Wikianon, Wikibase, Wikipelli, Will Beback Auto, William M. Connolley, Wingo, Winhunter, Wknight94, Wolfman, Wolfmankurd, Woohookitty, Woome, Wranglers 04, Wrcovington, Wscdfightyuim, Wtshymanski, Wyverald, X201, X570, Xaffect, Xaosflux, Xavier Combelle, Xelgen, Xevi, Xezbeth, XmDXtReMeK, Xpclient, Xrarey, XxXrah-chompXxX, Xyzzyplugh, Yamaguchi, Yamamoto Ichiro, Yaronf, Yashtulsyan, Ybbor, YellowMonkey, Yelyos, Yensin, Yngvarr, Yoganate79, Yonatan, Yosri, Yulius, Z3, ZachPruckowski, Zackmorris, Zanimum, Zanorath, Zanuga, Zarvok, Zastil, Zebbie, Zeldafreak104, Zemooo, Zman2000, Zoicon5, Zondor, Zzuuzz, Zzyzx11, Ævar Arnfjörð Bjarmason, Александър, Дарко Максимовић, , 2094 anonymous edits

Video_game *Source*: http://en.wikipedia.org/w/index.php?title=Video_game *Contributors*: -Majestic-, -jmac-, -the mole-, 041744, 04deacla, 11flanam, 12a21, 15dancing, 195.92.67.xxx, 1bobby93, 22vampsrock22, 25, 360sucksdick, 3m0ness, 7tattoosleeves7, 99DBSIMLR, A.J.S., ABF, ACSE, ADeveria, AMHR285, AVRS, AVeRY!, Aaaaqqqq, Aaronjosephcortez, Aarontay, Abby 82, Abce2, Academic Challenger, Acdx, Acidburn24m, Acroterion, AdAdAdAd, Adam Marx Squared, Adashiel, AdjustShift, Admrboltz, Adraeus, Adrian, Ae-a, Aeosynth, Aerielle 7@yahoo.com, AeronPeryton, Aespir, Afrosmokes, Ahmansoor, Ahmed193, Ahoerstemeier, Air bags, Airconswitch, Airplaneman, Aitias, Ajaxkroon, Ajbperc, Akaloc, Akeemcaffee, Aknorals,

Alai, Alansohn, Aldaron, Aldwin1, Ale jrb, Aleenf1, AlexiusHoratius, Alexkin, Alexnoman, Alexthe5th, AlistairMcMillan, Alksub, Altenmann, Altiego, Alucard (Dr.), Amilmoe, Amren, Anarchangel, Andonic, Andre Engels, Andrea105, Andrevan, Andrevruas, AndrewWTaylor, Andrewalot, Andromedabluesphere440, Andrwsc, Andy Marchbanks, Anetode, Angela2109, Angelo22spag, Angr, Angus Lepper, Animorphswes, Animum, Anna Frodesiak, Anna Lincoln, Ansalo, Antandrus, Antolage, Antonio Lopez, Apparition11, Appropriate Username, Aramee, Aranae, Arcanineghost, ArchStanton69, Armando12, ArnoLagrange, Arru, Artaxiad, Asbestos, Asher196, Ashley-hepton, Asterion, At09kg, At0micPix3l, Atb129, AtheWeatherman, Atif.t2, AuburnPilot, Auntof6, Autonova, Avenged Eightfold, Avoided, Awarmus, Awh, AzaToth, Azzy2233, BHC, Bachrach44, Ballinb15, Ballsakatk, Banes, Barek, BaronMaverick, Bart133, Bat on Fire, Bboessen, BcRIPster, Bearvomit, Beetstra, Beki710, Beland, Belasted, Bensin, Berenlazarus, Bergerj, Besieged, Betacommand, Beyond silence, Billie Pipier, Billinghurst, Billnyethesciencegirl, Bimbo1521, Biodronusmaximus, BirdKr, Bkonrad, Black Kite, Black Metal ist krieg, Blake-, Blanchardb, Blaxthos, Blehfu, Blue520, BlueEel, Bluezy, Bob f it, Bob facal, Bob the dee dee dee, Bobo192, Bobsag, BobtheVila, Bobtheamish, Bonadea, Bongwarrior, Bookandcoffee, Bookermorgan, Bookuser, Boomshadow, BorgQueen, Borisblue, BostonMA, Brad2534, Brandonthompson51, BrianGV, Bricearoo, Brighterorange, Broadmas, Broadmas214, Brock Miller, BrokenSegue, Bruguiea, Brutulf, Bryan Seecrets, Bsadowski1, Btilm, Buckeyefan105, Bulbous, Bulletproof50, BunnyDee, Bunyk, Burzmali, Businessman332211, Buuneko, Bwg789, CBecker, CIreland, CK2326, CRGreathouse, CT Cooper, CWY2190, CWenger, CWii, Cacapoop222, Cadsuane Melaidhrin, Calabraxthis, Callofduty007, Calmer Waters, Caltas, CalumH93, Cambrant, Camden town, Can't sleep, clown will eat me, Canadian-Bacon, CanadianLinuxUser, Canadianshoper, Canaima, Capricorn42, Card, CardinalDan, CaseyPenk, Catscratch, Caulde, Cdburnnner, Cek, Centrx, Cflm001, Chafiman, Chaosfeary, Charles Matthews, Charlierichmond, CharlotteWebb, Cheezerman, Cheezmon, Chenmas214, Chenxlee, Chickenyogurt, Chickyfuzz14, Chocolateboy, ChopMonkey, Chotisornmas214, Chris Heiner, Chris Rocen, Christian75, Christina Silverman, Chubbynucca, Cireshorty26, Cirt, Citicat, Cjhlib, Ck lostsword, Ckatz, Clamshell Deathtrap, Clarince63, Cleaner475, Cliff smith, Clues506, Clyde Miller, Cmputer, Cocoaguy, Codetiger, Cody03, Coleton900, Coll7, Cometstyles, Commander Shepard, CommonsDelinker, Comrade42, Connor1, Conroeactor, Conti, Conversion script, Coolcaesar, Corybechtold, Courcelles, CovertOps77, Cowman63, Cpavlovski, Craigrosa, Craigsjones, Cremepuff222, Crippsicle of tha nation, Crusader312, Cuchullain, Currymcelroy, Cyanoa Crylate, CyberSkull, Cyberdemon88, Cypie, Cyrus XIII, D, DACrowe10, DARTH SIDIOUS 2, DGG, DGaw, DH85868993, DMacks, DNS246, Da500063, DabMachine, Dageoffant, DaltinWentsworth, Dan D. Ric, Dan198792, DanMS, Danbur90, Dancter, Daniel2424, Danski14, DarkElf109, DarkLordSeth, DarkShroom, Darkn3ss416, Darren23, Darth Panda, Darthmauler1321, Dave101, Daveh4h, Daveydweeb, Davhorn, David Eppstein, David.Mestel, DavidHOzAu, Davidizer13, Davidshq, Daymas214, Dcarriker92, DeadMansFilm, Deathsin366, Debresser, Decltype, Dekimasu, Delirium, Dell0154, Delldot, Delta 51, Demonkey36, Den fjättrade ankan, Denelson83, DerHexer, Descent, Desertrose0601, Dg0896, Dgw, Dhill5427, Diannaa, Diego Moya, DietCaffeineFreeCherryCoke, Digdug41, Dillongp, Dinoguy2, Discospinster, Dispenser, Dlohcierekim, Dlohcierekim's sock, Docski35, Doenherleven, Dogman15, Dogzilly150, Doomedtx, Doominator, Doug Bell, Doyley, Dpeters11, Dputig07, Dr Doomed, Dr-fernman, Dr.Bhatta, DragonBlazer57, Dreambeaver, Dreammchine64, Drew's Friend, Drick99, Drmies, Drumpler, Dshallard, Dude323, Dudethecat, Dungodung, Duozmo, Dureo, DurotarLord, DurtyWilly, Dv82matt, Dwayne, Dyaa, Dycedarg, Dynesclan, Dysepsion, E. Ripley, E946, ENeville, ESkog, EStew1234, EamonnPKeane, East718, Eberk456, EddieSegoura, Eden Thinker, Edgar181, Edward, Eeekster, Eep^2, Eequor, Effer, Egmontaz, Ehend661, Ehheh, El Krem, ElectricEye, Electricat, Eleph, Emuzesto, Endersoul, Energetic is francine@yahoo.com, Entropy, Enviroboy, Enzarion, Epbr123, Eragon 10, Erianna, Ert123101, Escape Orbit, Escort 89, Estarriol, Euchiasmus, Eurosong, Euyyn, Evansanity, Evaristé93, Evercat, Everyking, Evilasian3, Ewlyahoocom, Excirial, Exothermic Reaction, Extransit, Eyu100, F.Pavkovic, FCYTravis, FFF 4 me, FMAFan1990, Fabledrac, Fairfieldfencer, Falcon8765, Fan-1967, Fang Aili, Farner777, Favonian, Fdm20cgsiegel, FelisLeo, Felsir, Feltonmas214, Femto, Fenix99, Finalius, Finduilas 09, Fisherman Vs. Ghostface, Flatterworld, Fleinn, Fleminra, Flubeca, Flyingimc23, Foland, Fomz, For great justice., Force92i, Fortdj33, Fourth hour fun, FrankSeguin, Frap, Frecklefoot, FrederickMS, Fredrik, FreeMorpheme, Freez123, FreplySpang, Freshacconci, Froid, FrostyBytes, Froth, Funandtrvl, Funeral, Funpcgame, Fyyer, Fæ, G.W., G4rfunkel, GRBerry, GTBacchus, Gabbe, GabrielF, Gaius Cornelius, Galactor213, GameLegend, Gamerfm maxc, GamesGuru, Gamester501, Gareth Aus, Gary King, Gasheadau, Gdo01, Geezergaz77, Geni, Georgewilliamherbert, Geralt92, Gerg131, GermanX, Ghasforty, GhostPirate, GhostRobo, Giggy, Gillette.ben, Gilliam, Gingivitisxlg, Ginkgo100, Ginsengbomb, Gioto, Glacialfox, Glane23, GlassCobra, Glenn, Glob.au, Goatsetubgirl, GodMan, Golbez, Gomm, Gonzo fan2007, Good Olfactory, Goodbye Galaxy, Goodvac, Gookey, GorillaWarfare, GothicPanda, GraemeL, GrafZahl, Graham87, Gravitan, Grayshi, Greenrd, Greeves, Gregfitzy, Gregorik, Gremlin Gremlin, Grey Shadow, Greyengine5, Greyhead, Gsarwa, Gscshoyru, Gtrmp, Gundamforce, Gurch, Guyinblack25, Gwernol, HALOMISTRO, HJ Mitchell, HMP22, Hadal, Hahnchen, Haipa Doragon, Hairy, HairyPerry, Hall Monitor, Halo master chief, Happypie666, Harland1, Harmo777, Harro5, Harryboyles, Hasek is the best, Haseo9999, Hawaiian717, Hazzerd3, Hdt83, Headbomb, HeiRenXuesheng, Hell Pé, Hellcat fighter, Hellknowz, Hello1792, Heltec, Henry Gray, Hetzer, Hevernon, HiDrNick, Hibana, Hiddekel, Hiddenfromview, Highcount, Hihihixxx, Hik4ru123, Hinubs2, Hirokazu, Hitman12319, HjacksonvVT, Hmrox, Holek, Hollaholla, Holmesey, HonztheBusDriver, Horst F JENS, Housethe4, Hu12, Huck123555, Hunter1084, Husnock, Husond, Hut 8.5, Hysterionproteus, HyuenK, I am a violinist, I3lizzard, IMNOTAVANDALK5, IRP, IXella007, Iamryan365, Ian Pitchford, Ian.thomson, Ibai123, Iced Kola, Icedoghanson77, Icestorm815, Ichangewords, Ichormosquito, Igordebraga, Ihateyou2544261, Ikryn, Ilikemusic, Immunize, Imnotminkus, Imroy, Independent Journalist, Independentfilmactor, Indrian, Instinct, InterTechnologies, Interiot, Interrobamf, Ipwnyounub, Iridescent, Irishguy, IronGargoyle, Ishikawa Minoru, Ivan530, Ixfd64, Izzy007, J fantastic, J-stan, J.L.Main, J.delanoy, J04n, J167201, JForget, JNW, JPG-GR, JRGBruno, JYi, Ja 62, Jacian7, Jack Espe, Jacob Finn, Jacoplane, Jacroe, Jagged 85, Jahiegel, Jaman61994, Jamesday, Jamesmas204, Jamesontai, Jappalang, Jaranda, Jarble, Jasonauk, Jauerback, JauhienIJ, Jaxl, JayC, Jbinder, Jc37, Jeanpuetz, Jeeny, Jeff G., Jeff7, Jellyvista, Jemijohn, JeremyA, Jerichi, JesseW, Jet57, Jfpierce, Jhoony, Jilliansgarner, Jim1138, Jimothytrotter, Jitse Niesen, Jklin, Jmanigold, Jmcfarlan, Jmh649, Jmlk17, Jmoney2323, JoanneB, JoeLevel, JoeSmack, John Fader, John Wilcox, John smith 99, John254, JohnCD, JohnnyB256, JohnnyMrNinja, Joie de Vivre, Jonodabomb, Jordiehowe, Joseph Collins, Josh Grosse, Josh Parris, Josh1789, Josquius, Jossi, Jovianeye, Joyous!, Jpgordon, Jreconomy, Jtalledo, JuJube, Jubeidono, Juiceboxer, Juliancolton, Julianmurphy88, Jununhere, Jusdafax, Just James, Justadude4, Jwhimmelspach, Jóna Þórunn, K1Bond007, KBi, KNHaw, Kablammo, Kaeptn1, Kakashi hatake76, Kansan, Karam.Anthony.K, Kareeser, Karimiwiki, Kariteh, Karlscherer3, Kate Phaye, Katxijasotzaile, Kayau, Kbh3rd, Kchishol1970, Kcwoo, Keegan, Keilana, Kel-nage, Kelly Martin, Kelvinc, Kennyannydenny, Kerotan, KerryO77, Kester Teague, Keveth, Khanmas214, Khfan93, KieferFL, KieferSkunk, KiloByte, Kimse, Kinema, King Bob324, King Lopez, KingR20, KingTT, Kingpin13, Kingzy1, Kittins floating in the sky yay, Kjkjava, Klevenskiy, Klondike1098, Kmg90, Kms, KnowBuddy, KnowledgeOfSelf, Koavf, Kookylad, Kozuch, Kozy44, Kpolsson, KrakatoaKatie, KramarDanIkabu, Krash, Krellis, Kstrones, Ksyrie, Kuguar03, Kukini, Kuru, Kurykh, Kyle Barbour, Kyle key, KyuuA4, L Kensington, LAW, LFaraone, LOL, Lady Jaye, Lakota5, Lalalilililololi, Lammas214, Landroo, LarsBB, Larsinio, Latka, Lavenderi, Leafyplant, LearningKnight, LeaveSleaves, LedgendGamer, Leet Sher, Legoboy1129, LemonPuddingPop, LeoNomis, Leszek Jańczuk, Letssee23, Levineps, Lightmouse, Lights, Lijakaca, Lildrumboy5000, Literacola, Little Mountain 5, LittleOldMe, Littleclaude, Llewdor, Lmartop, Lockeownzj00, Logan, Logan Rice, Lolman1341, Look123, LordShitzu, Lordmontu, LtNOWIS, Lthandley, Lthornsb, Lucasbballplaya, LuckyBob76, Luisedgarf, Luna Santin, Lupin, MAXXX-309, MC Scared of Bees, MER-C, MIT Trekkie, MK8, MONGO, MPN, MTN, MVNdude, Mabwallace, MachoFern, Macintosh User, Mad gamers, Mad031683, Maddie!, Madnesstosadness, Magicnxnja, Magioladitis, Magister Mathematicae, Magog the Ogre, Mahjongg, Mahyar d zig, Mainak, Majorly, Make, ManPingas, Mangojuice, Maralia, Marasmusine, Marechal Ney, Marek69, Mario 4, Marioromeroaguirre, Mark Richards, MarkBuckles, MarritzN, Martarius, Martin451, Marvelvsdc, Maser Fletcher, Massi889, Master Bigode, Master shepherd, MasterCory, MasterOfHisOwnDomain, Masterm900, Materialscientist, Mathewbeker, Matt Britt, Matt Neuteboom, Matteh, Mattguy2k, Matthew Auger, Mattkloz, Maueze, Maxim, Mboverload, McGeddon, Mckaysalisbury, Mdebets, Meelar, Meggar, Mel Etitis, Melodia, Meme133ere, Mempeni09, Menshoes, Mentifisto, MentoringBoys, MetsFan76, Meursault2004, Mg2594, Mhanimemastr, MiLo28, Michael Devore, MichaelBillington, Michaelpb, Mike Rosoft, Mikeo, Milenkovic214, MinderOne, Mindtrap900, Minimac, MisterHand, Misza13, Mitaphane, Mitsukai, Mnrubin, Molerner, Mondocanerules, Monkey4christ76, Montwel03, Moogwrench, Moomoomoo, Moped, Motherphucker, Moxiemag, Mqduck, Mr Cool is here, Mr. Bobbleweed, Mr. Strong Bad, Mr. XYZ, MrBlueSky, MrDaBucket, MrDrak, MrFish, MrOllie, MrSomeone, MrWallet, MrWhich, Mrcolj, Mrees1997, Mrwojo, Ms2ger, Mschel, Mspraveen, Muchness, MudkipNDS, Mushroom, Muttdog, Mwanner, Mwongozi, Mypointusingyoutube, Mysterius, Mythdon, Mytildebang, Mywikiaccount47, N. Harmonik, NAHID, NERIC-Security, NGPriest, NHRHS2010, Nadez1234, Nakon, Nate0ate, NateDan, NawlinWiki, Nburden, Nczempin, NerdyScienceDude, Nessy2100, Neverquick, NewEnglandYankee, Newb360, Newmansan, Nexic, Nick, Nick R, Nifboy, Nigholith, Nightwolf1987, Nihiltres, Nikai, Nikitko, Nintendo Power Magazine Reader Boy, Nitesh512, Nitya Dharma, Nivix, Nlu, NoSeptember, Noozgroop, Notinasnaid, Nrodballer13, Nstrud, Nudecline, Nufy8, Nuno Tavares, Nyfinestswagger2, Nzrouillard, OMGWTFBBQ???(2), Oberious, Oblivious, Octet7, Octoberhero, Ohnoitsjamie, Olaan, Old Moonraker, Oli Filth, Oliver202, Ollydv, Ollyollyster, Omgthatpie, Omicronpersei8, One, Ongar the World-Weary, Oni Ookami Alfador, Oobnoob, Optichan, Orangemango, OrgasGirl, Ost316, Ottershrew, Out slide, OwenX, Oxymoron83, PIrish, PMG123, Paada, Pagrashtak, Pak21, Papillon981, Parapedia, Parkjunwung, Parrothead1983, Particlepeople, Patdreams, PaulGarner, Paulcmnt, Pearle, Pek, Pengo, Pennywisdom2099, Pepper, PerEdman, PeregrineAY, Persian Poet Gal, Peruvianllama, Pete G., PeterEastern, PeterSymonds, Pevernagie, Pgk, Ph3z, Phdrummer, Phgao, Phil Boswell, PhilMacD, Phildogg82, Philip Trueman, Philip.t.day, PhilipO, Philippe, Phoenix1177, Phoinex18, Phsycoawesomeness, Piano non troppo, Pieleric, Pikachu025, Pikawil, Pingular, Pinikas, Pj.220, Plaguex1, Planetary Chaos, Playclever, Playstationdude, Plesiosaur, PlueXD, Pluma, PokerPlayer, Pomegranate, Pomte, Porterjoh, Potatoe292, Potlasubodh, Power level (Dragon Ball), Poxmarc, Ppk01, Prak, Prashanthns, Premiereboris, PrestonH, PrincessofLlyr, Profquimby, Proofreader77, Proteus, Pseudo Intellectual, PseudoOne, Psicorps, Puchiko, Pumas214, Puninator m12, Pupster21, Purplepox01, Quadell, Quantumobserver, Quarl, Quasipalm, Queen Rhana, Quinsareth, Quintote, Quoth, Qxz, R'n'B, RG2, RJFJR, RMHED, RTC, Raconteurs, Radagast83, Radon210, RagingR2, Ram4eva, Rama, Ran4, Randall Brackett, RandomStringOfCharacters, Ratherhaveaheart, Raven4x4x, Raveparty20, Raylu, Rdsmith4, Reach Out to the Truth, Recury, Red, Red Director, RedHillian, Redattack34, Redthoreau, Redvers, Reedy, Reekan12, RegainTheTruth, Regancy42, RelishingVictor, Remnar, Remurmur, Renodivorce, Requestion, Res2216firestar, RexMaztar, RexNL, ReyBrujo, Rgc, Rhobite, Rich Farmbrough, Ridernyc, Riley dix, Rishikessivaonly1, Rising*From*Ashes, Rjd0060, Rjwilmsi, Rkraider, Rlk89, Rmb82, Rob T Firefly, Robert K S, Robert P. O'Shea, Robertsbrent, Robin Patterson, Robth, Rocafella, RockMFR, Rodrigo Cornejo, Roger.smith, Romney yw, Ron Howard, Ronhjones, Rory O'Kane, Rotard4989, Roux, RoyBoy, RoyLaurie, Rrburke, Rtyq2, Rubyrocks, Ruggert1, Rwwww, RyanParis, Ryanjunk, Ryulong, S h i v a (Visnu), SB Johnny, SDC, SJP, SSerggg, ST47, SYSS Mouse, Sade, SalomonMuriel, Salty!, Salvio giuliano, Sam Blacketer, Sam Hocevar, Samashton, Samebchase, Samsara, Samuraispy, Samwb123, Sander123, Santer, Scaife, Scandum, Scarian, Scepia, Schmidtmandaddy, Scientizzle, Scissorhead335, Scottymoze, Scramble, Sdornan, SeOatriX, Sean William, Seaphoto, Searchme, Seba5618, SeeSchloss, Seelys, Seinxxvi, Seraphimblade, Sercimas214, Serious Mr. Karate, Serj75, Sesshomaru, Sevela.p, SeventhTrilogy, SevereTireDamage, Seviegold248, Sexykins13, Sfdecker, Shanes, SharkD, Sharon08tam, ShaunES, Shawnc, Shawnhath, ShelfSkewed, Shirimasen, Shiroi Hane, Shizhao, Shoaler, Shoeofdeath, Shotwell, Shpriggs, Shuipzv3, ShurikenNaruto, Sigma 7, Silly rabbit, Silver Edge, Silverrevlis, Simesa, SimonTrew, SimpleName, SinglishPerson, Singularity, Singularity42, Sippan, Sirgregmac, Sisyphustkd, Sitearm, SjaichudinMAS214, Sjv27, Skizzik, Skullsbud, SkyWalker, Skybon, Slady, Slike, Slike2, Slon02, Slumbergirls101, Smilesfozwood, Snakeyes (usurped), SnappingTurtle, Snaxorb, Snori, Snowolf, Soawsome, Sobolewski, Soetermans, Sohmc, Someonesdad363616, Sometimessubtle, Sonicdudeish, Sony Ericsson W950, Soonhuat95, Soopies, Sopoforic, Sorn67, Soumyasch, SpaceFlight89, Spark331, Sparky the Seventh Chaos, SparrowsWing, Spellcast, Spellmaster, Spike Wilbury, Spikesun2007, SpuriousQ, Sqkvii, Squid1996, St3w-101, Stan Shebs, Stanko, StaticGull, StefanDasbach, Stephenb, Stereotek, Stewartadcock, Stickauto, Storkk, Stormie, Strabismus, Striderdan, Sturm br, Stusutcliffe, Stwalkerster, Suckaz, Sumsum2010, SupaSoldier, Super-Earth, SuperDude115, Suttipongkanasaimas214, Svetovid, SwisterTwister, Symane, SynergyBlades, Synthe, Synthetik, Syrthiss, T-Diz, T3h videogamer!!!, TEX MrPistol, THEN WHO WAS PHONE?, THollan, TKD, TUMAS214, TWINKY52, Tabletop, Tachyon01, Tan canada, Tangense, Tangerines, Tangmas214, Tangotango, Tapir Terrific, Tarff, Targumon, Tarinth, TarkusAB, Taroaldo, TastyPoutine, Tauraunt, Teddyii0, Tedius Zanarukando, Teggles, Teh one, Tehcoolzar, Tellyaddict, Tempshill, Tendulkarman, Tenguman, TennisPl8r, Terenn Blackfield, Terrx, Tetraedycal, TexMurphy, Texture, Tezero, Tgmonkeyman, Th1rt3en, Th3 pr0gram, Thatguyflint, The Fwanksta, The Hybrid, The Moose, The Prince of Darkness, The Rambling Man, The Red Baron, The Rogue Penguin, The Thing That Should Not Be, The Transhumanist, TheFarix, TheKMan, TheMadBaron, TheMidnighters, TheMonkofDestiny, TheRanger, Thecrocker, Theda, Thehelpfulone, TheloniusMaximus, Theoneintraining, Thepcnerd,

Thespankguy, Thewallowmaker, Thibbs, Thingg, Thiseye, Thisisurmomma69, Thomas Larsen, Thomas Matthews, Thorek329, Thorpe, ThreeAnswers, Thu, Tiaccait, Ticknoris, Tide rolls, TigerShark, Timmy mctimmy, Tiptoety, Titoxd, Tjoeb123, Tobby72, TomStar81, Tomgdoe16, Tommosimmo, Tommy2010, Tomsteele, Toniof657, Tonku, Toredid, Tot12, Touch Of Light, Tr00st, Tree Biting Conspiracy, Tregoweth, Tresiden, Trick man01, Tridonias, Tron09, Tronno, Troutmand, Truedat323, Trusilver, Tubby123456789, Tuckman8, Tulkolahten, Turnipface, TutterMouse, Twang, TwinnedChimera, Twinsday, Typer 525, Typhin, Tyranic-moron, Tyros1972, UlitmateLoserNo1, Ulric1313, Ultimus, Umother, Uncle Dick, Uncle G, Urena198, Useight, Useless93, Username314, Utcursch, VAcharon, VPeric, Vanished User 1004, Vdrakon, Veinor, Vendettax, Versageek, Versus22, Vgranucci, Videogamer128, Videogamessuck, Vigorousjammer, Violetriga, Vishnava, Visor, Vizjim, Vla.67, Voidvector, Voliotis1453, Voltron, Voretus, Voyaging, Vp???, Vranak, Vrenator, Vshiz2k7, Vydran, W3stfa11, WOSlinker, Waggers, Wahrmund, Waldir, Wampuscat nene, Warboi, Wavelength, Waywardhorizons, Wbm1058, Weasil102, Wedgeh, Weeblewobbles, Weefz, Weinerheadshaboom, Well, girl, look at you!, Wgungfu, Whalen207, Where, White Shadows, WhiteCat, Whitepaw, WiiStation36013digits, Wiki Roxor, Wiki alf, WikiLaurent, WikiWha?, Wikieditor06, Wikipedical, Willbrydo, Williamamy, Williebruciestewie, Willking1979, Wilsonbiggs, Wimt, Wizardman, Wknight94, Wlievens, Wolfrider, Woodleynye, WookieInHeat, Wordofmouse, Wossi, Wowowwiki, Wps2010, Wsvlqc, Wtmitchell, XJamRastafire, XanTaa, Xaphnir, Xeno, Xenoploid, Xenoveritas, Xezbeth, XopieX20, Xrhetor, Y0u, YORD-the-unknown, YVNP, Yamaka122, Yamamoto Ichiro, Yerpo, Yeti Brothers, Yipes! Its a gamer!, Ymirfrostgiant, Yobb, Yoenit, Yosef1987, YourEyesOnly, Ypacaraí, YvonneLe, ZS, ZXCVBNMQAZ, Zachary333, Zahid Abdassabur, Zapexe, Zarek, Zavv06, ZeWrestler, Zeek989, Zelda fan51, Zeldafanjtl, Zenohockey, Zmea1, Zoganes, Zollerriia, Ztp912, Zxcvbnm, Zzuuzz, Лев Дубовой, קוראפ, 3484 anonymous edits

Football Manager *Source*: http://en.wikipedia.org/w/index.php?title=Football_Manager *Contributors*: -Majestic-, 07ed01, AMsisscorsister, Acerperi, Adam4267, Akrabbim, Alias Flood, Alynna Kasmira, Amunra reborn, Andybazslim, Anonymous101, Ashl, Auntof6, Barnolde, Beetstra, Bencey, Benson85, Bluezy, Boothy443, Bornintheguz, Brianreading, Bubba hotep, Buttons to Push Buttons, CambridgeBayWeather, CanisRufus, CapPixel, Capricorn42, Carioca, Cdc, Chairboy, Chamal N, Charlie Allen, Cheat2win, Chico75, Chowbok, Chris the Luddite, Clifton137, Cometstyles, CyberSkull, Cyclonenim, Cyoshio, Cyraj, Cyrius, D.brodale, DE, DJ Clayworth, Dale Arnett, Dan Atkinson, DanGravell, Darthvadar92, Db1987db, DebiQue, Deconstructhis, Dgkdk, Discospinster, Djdannyp, Djungelurban, Djx999, Dmaclach, Dreaded Walrus, DrunkenSmurf, Dudesleeper, Dukeseee, ESkog, Earl CG, Editor at Large, El Pantera, El salvador, Elmer Clark, Estel, Evil Monkey, Falcon9x5, Firsfron, Fish1989, Fmportal, Freezedriedpop, Frodet, Frog19021996, GPattle, GangstaEB, Gareth E Kegg, Gary King, General garcia, George14789, George747, GiantSnowman, Gleb Belov, GoingBatty, Gravitan, Greek Transistor, Groove1984, Grue, Hagibest, Heightwatcher, Hien712, Hunocsi, Husond, Infrogmation, Inkington, J Milburn, J.delanoy, Janadore, Jankuza, JauNe, Jce0025, Jeff3000, Jeffmister, JeremyA, Jimmyc85, JoeSmack, JohnnyPolo24, Jonnif, Joraar, Joy, Jwkpiano1, Kevin McE, Kiand, Kingjeff, Kingutd, Kongen af fodbold, Krawi, Ksyrie, Lawlore, Leedsutd95, Lewispb, Lihaas, Lord Bob, Lupin, Malcjennings, Maniacgeorge, Manop, Marasmusine, Marek69, Mariner3, Martial75, Master Thief Garrett, Megata Sanshiro, Mikadk, MikeUpNorth, Millm0w, Mitchan, Mitea, Mlaffs, MrKIA11, NES Wii, NSLE, Nathancorlett, NellieBly, Nemodomi, Nephersir7, Nk, Nufc2006, Nv8200p, Ny156uk, PMLF, Pal, PeeJay2K3, Pfortuny, Phantom Thief, Piano non troppo, Pirlo23, Qwghlm, Qwyrxian, R'n'B, R107, RandomStringOfCharacters, Recury, Requestion, Rfl, Rockpocket, Rosoner, Rtic, Ruthlessrob, S3000, Salishsea, Sam Korn, Sammybrowne123, Sassf, Scfcpaddy, Shell Kinney, Shi03066, Silvermane, Slangius, Sykko, TJ Spyke, Tabletop, TasmaniaUnitedFan, Tdrawler, The Thing That Should Not Be, Thorpe, Thrapper, Thumperward, Tide rolls, Tim1357, Timc, Timlarsen13, Toon05, Torstein, Ugen64, Ugur Basak, Ukcreation, Ukdan999, Utcursch, Viper1288, Welsh, Wezz, Will 3rd, Willking1979, WolfenSilva, X201, XSEARL3Yx, Yankees76, Yanksox, Yono, Yonthe, Zakey, Zephyr1981, ZerBox, Zocky, Ålesund, 686 anonymous edits

Championship Manager *Source*: http://en.wikipedia.org/w/index.php?title=Championship_Manager *Contributors*: Akrabbim, Bkell, Cander0000, Chensiyuan, Djungelurban, Dudesleeper, Friday, GamaFranco, George2001hi, H4stings, Harthacnut, Hrcolyer, Iamthouth, JIP, Lihaas, Megata Sanshiro, MichaelMaggs, Mika1h, N. Harmonik, Nightrider 83, Pablo323, Quoth, Salavat, Sjc196, That Guy, From That Show!, Ukdan999, WikiManuel, Winstonhyypia, 32 anonymous edits

Sports_Interactive *Source*: http://en.wikipedia.org/w/index.php?title=Sports_Interactive *Contributors*: Aaaaeh?, Ahoerstemeier, Akrabbim, Andrew Levine, Barrien, Carioca, Celldweller, Cube b3, DMG413, Dan Atkinson, DanGravell, Divine time, Doktor Wilhelm, Dreaded Walrus, ENG422, Elonka, Everyking, Fakeymcname, Fledgeling, Flipkin, Ganryuu, Geni, Gogo Dodo, Grue, Hcheney, HexaChord, Hig Hertenfleurst, Inseverance, Inuboy1000, J.delanoy, JHunterJ, Janadore, Jean-Frédéric, Jimfbleak, Jontintinjordan, Kerotan, Kesac, KingTT, Kingutd, Kongen af fodbold, Lenzar, LilHelpa, Mariner3, Mark272, Martarius, Megata Sanshiro, Michael Snow, Moink, Pcpcpc, PeeJay2K3, RadioKirk, Red Phoenix, Redefining form, RickK, Rjwilmsi, Ronning, Skaterdude182, Sportman2, Steve.battisti, Stewartadcock, Tabletop, Thumperward, Ukdan999, X201, 146 anonymous edits

Front_Office_Football *Source*: http://en.wikipedia.org/w/index.php?title=Front_Office_Football *Contributors*: Achangeisasgoodasa, Amalas, Anationofmillions, Babydaddy, Barkeep49, Bongwarrior, Ciperl, ClockworkLunch, Cthomer5000, Daniel, Discospinster, Domonas, Dr31, Fuhghettaboutit, Gaius Cornelius, GregorB, HJ Mitchell, Hell Atlantic, JAF1970, Kingof7, Lorson, Mack2, Nick, Paleorthid, Pax:Vobiscum, Redwolf2010, RexNL, Tbhotch, TheParanoidOne, TnS, TredWel, Zzyzx11, 76 anonymous edits

Lemonade Stand *Source*: http://en.wikipedia.org/w/index.php?title=Lemonade_Stand *Contributors*: 23skidoo, Aaronbrick, Bart133, Bunchofgrapes, DanFein, Davemcarlson, Deepred6502, Dougk, ESkog, Ejalbert, Estephan500, Explicit, Feezo, FourteenDays, Frecklefoot, Fullobeans, GVnayR, H3xx, Jeffrey Henning, Kcowolf, King of Hearts, Ktdreyer, Kurivaim, Lizardo tx, MONGO, Mansuetodigital, Micru, Molerat, MrCEO, MrOllie, Nasa-verve, Nubby, Rhetth, Rjwilmsi, RobJ1981, Robert Weemeyer, Sabrewing, Surv1v4l1st, Thumperward, Useight, Wikidudeman, Wikievan9, Yorick8080, Zbegner000, 46 anonymous edits

Government_simulation_game *Source*: http://en.wikipedia.org/w/index.php?title=Government_simulation_game *Contributors*: 7704386800, Academic Challenger, Adilsonmrj, Aequo, Angeldk 6, AxelBoldt, Bcarlson33, Benshelmars, Bethpage89, Blandish, Blue403, Bonadea, BostonJack, Bragador, Brandonazz, Briannine, Bridies, Bruce Tedder, Cbrown1023, Centred, Chabuk, Chris Latour, Chris657, ChrisGlew, Chrisglew, Citizen Sykes, Colonies Chris, Coolsirch88, Crosland, DJ Clayworth, DVdm, Dancrosoft, Danthemankhan, Dark Serge, Davidmagnusalexander, Deathby, Devari, Diderot's Ghost, Dusoft, Eastlaw, Elryacko, Eterry, Falcon8765, Func, Gaius Cornelius, Hohum, Hoygan!!, Hypnosifl, Imamathwiz, Impact red, Indoles, Inter, Ismail, Itake, Iune, Jagged 85, JamesBWatson, Jammydodger, Jsmorse47, Kaa, Katydidit, Keeperoftheseal, KevinOKeeffe, Kewpid, Kizor, Kozuch, Kross, Leuko, Lloegr-Cymru, M kwiat1974, Mailer diablo, Malone70, Mandarax, Margrave, Mark Renier, McGeddon, Meeso, Mike Rosoft, Mike.lifeguard, Miqademus, Mr GreenOrange15, Mr. Stradivarius, MrOllie, Mummer, N. Harmonik, NSR, NW036, Nae'blis, Narcisse, NavalWar1800, Niceguyedc, Nickjbor, Noaccountaccount, Northamerica1000, Notsorockstar, Oberst, Okeeffedk, Ost316, Peer-LAN, Percy Snoodle, Peyre, Pinkkeith, Plasticup, Plrk, Randomran, RexNL, RockMFR, RoySmith, Saerain, SanityFreeZone, Sasha Slutsker, Schmidtr1, Scott McTavish, Sfan00 IMG, SharkD, ShelfSkewed, Shii, Someone another, Sovietmessiah, Spotle, Stevietheman, Sunray, Tarinth, Thrasherssyn3443, Tntoak, Tracer9999, Tyking, VICPrem, Wlievens, Woohookitty, Wprlh, Wtmitchell, Wyatt Riot, XGhost4000x, Yamamoto Ichiro, 335 anonymous edits

Image Sources, Licenses and Contributors

File:DM IBM S360.jpg *Source*: http://en.wikipedia.org/w/index.php?title=File:DM_IBM_S360.jpg *License*: unknown *Contributors*: Ben Franske

Image:Columbia Supercomputer - NASA Advanced Supercomputing Facility.jpg *Source*: http://en.wikipedia.org/w/index.php?title=File:Columbia_Supercomputer_-_NASA_Advanced_Supercomputing_Facility.jpg *License*: unknown *Contributors*: Trower, NASA

Image:Intertec Superbrain.jpg *Source*: http://en.wikipedia.org/w/index.php?title=File:Intertec_Superbrain.jpg *License*: unknown *Contributors*: Brighterorange, 1 anonymous edits

File:2010-01-26-technikkrempel-by-RalfR-05.jpg *Source*: http://en.wikipedia.org/w/index.php?title=File:2010-01-26-technikkrempel-by-RalfR-05.jpg *License*: unknown *Contributors*: User:Ralf Roletschek

File:Thinking Machines Connection Machine CM-5 Frostburg 2.jpg *Source*: http://en.wikipedia.org/w/index.php?title=File:Thinking_Machines_Connection_Machine_CM-5_Frostburg_2.jpg *License*: unknown *Contributors*: Mark Pellegrini

File:G5 supplying Wikipedia via Gigabit at the Lange Nacht der Wissenschaften 2006 in Dresden.JPG *Source*: http://en.wikipedia.org/w/index.php?title=File:G5_supplying_Wikipedia_via_Gigabit_at_the_Lange_Nacht_der_Wissenschaften_2006_in_Dresden.JPG *License*: unknown *Contributors*: Conrad Nutschan

File:Acer Aspire 8920 Gemstone by Georgy.JPG *Source*: http://en.wikipedia.org/w/index.php?title=File:Acer_Aspire_8920_Gemstone_by_Georgy.JPG *License*: unknown *Contributors*: User:Georgy90

File:Acorn BBC Master Series Microcomputer.jpg *Source*: http://en.wikipedia.org/w/index.php?title=File:Acorn_BBC_Master_Series_Microcomputer.jpg *License*: unknown *Contributors*: MarkusHagenlocher, Mono, Ubcule

File:Dell PowerEdge Servers.jpg *Source*: http://en.wikipedia.org/w/index.php?title=File:Dell_PowerEdge_Servers.jpg *License*: unknown *Contributors*: User:Dsv

File:Jacquard.loom.full.view.jpg *Source*: http://en.wikipedia.org/w/index.php?title=File:Jacquard.loom.full.view.jpg *License*: unknown *Contributors*: User Ghw on en.wikipedia

File:Jacquard Joseph Marie woven silk.jpg *Source*: http://en.wikipedia.org/w/index.php?title=File:Jacquard_Joseph_Marie_woven_silk.jpg *License*: unknown *Contributors*: Astrochemist, Ecummenic, Ezrdr, Kilom691, WikipediaMaster

File:Z3 Deutsches Museum.JPG *Source*: http://en.wikipedia.org/w/index.php?title=File:Z3_Deutsches_Museum.JPG *License*: unknown *Contributors*: Original uploader was Venusianer at de.wikipedia (Original text : Venusianer 14:13, 3. Jan. 2007 (CET))

File:Classic shot of the ENIAC.jpg *Source*: http://en.wikipedia.org/w/index.php?title=File:Classic_shot_of_the_ENIAC.jpg *License*: unknown *Contributors*: Unidentified U.S. Army photographer

File:EDSAC (10).jpg *Source*: http://en.wikipedia.org/w/index.php?title=File:EDSAC_(10).jpg *License*: unknown *Contributors*: Copyright Computer Laboratory, University of Cambridge. Reproduced by permission.

File:80486dx2-large.jpg *Source*: http://en.wikipedia.org/w/index.php?title=File:80486dx2-large.jpg *License*: unknown *Contributors*: A23cd-s, Adambro, Admrboltz, Artnnerisa, CarolSpears, Denniss, Greudin, Julia W, Kozuch, Martin Kozák, Mattbuck, Rjd0060, Rocket000, 11 anonymous edits

File:SSEM Manchester museum.jpg *Source*: http://en.wikipedia.org/w/index.php?title=File:SSEM_Manchester_museum.jpg *License*: unknown *Contributors*: User:Parrot of Doom

File:H96566k.jpg *Source*: http://en.wikipedia.org/w/index.php?title=File:H96566k.jpg *License*: unknown *Contributors*: Courtesy of the Naval Surface Warfare Center, Dahlgren, VA., 1988.

File:FortranCardPROJ039.agr.jpg *Source*: http://en.wikipedia.org/w/index.php?title=File:FortranCardPROJ039.agr.jpg *License*: unknown *Contributors*: Arnold Reinhold

File:Mips32 addi.svg *Source*: http://en.wikipedia.org/w/index.php?title=File:Mips32_addi.svg *License*: unknown *Contributors*: German, JMCC1, Nachcommonsverschieber

File:Magnetic core.jpg *Source*: http://en.wikipedia.org/w/index.php?title=File:Magnetic_core.jpg *License*: unknown *Contributors*: Apalsola, Fayenatic london, Gribozavr, Uberpenguin

File:HDDspin.JPG *Source*: http://en.wikipedia.org/w/index.php?title=File:HDDspin.JPG *License*: unknown *Contributors*: Alpha six from Germany

File:Cray 2 Arts et Metiers dsc03940.jpg *Source*: http://en.wikipedia.org/w/index.php?title=File:Cray_2_Arts_et_Metiers_dsc03940.jpg *License*: unknown *Contributors*: User:David.Monniaux

File:Internet map 1024.jpg *Source*: http://en.wikipedia.org/w/index.php?title=File:Internet_map_1024.jpg *License*: unknown *Contributors*: Barrett Lyon The Opte Project

File:Nolan Bushnell - Game Developers Conference Online 2011 (4).jpg *Source*: http://en.wikipedia.org/w/index.php?title=File:Nolan_Bushnell_-_Game_Developers_Conference_Online_2011_(4).jpg *License*: unknown *Contributors*: Max Photography for GDC Online

File:Virtual-camera-system.png *Source*: http://en.wikipedia.org/w/index.php?title=File:Virtual-camera-system.png *License*: unknown *Contributors*: User:WikiLaurent

File:UMK3 Fighter Glitch.png *Source*: http://en.wikipedia.org/w/index.php?title=File:UMK3_Fighter_Glitch.png *License*: unknown *Contributors*: Alansohn, Khiy, Master Bigode, N. Harmonik, Tbhotch, 3 anonymous edits

File:Videogameretaildisplay.jpg *Source*: http://en.wikipedia.org/w/index.php?title=File:Videogameretaildisplay.jpg *License*: unknown *Contributors*: Bayo, Holek, Notwist

Image:Championship Manager Coverart.png *Source*: http://en.wikipedia.org/w/index.php?title=File:Championship_Manager_Coverart.png *License*: unknown *Contributors*: AvicAWB, Salavat, X201

Image:Sports Interactive badge.jpg *Source*: http://en.wikipedia.org/w/index.php?title=File:Sports_Interactive_badge.jpg *License*: unknown *Contributors*: Grue, X201

File:Diplomacy pcgame screenshot.jpg *Source*: http://en.wikipedia.org/w/index.php?title=File:Diplomacy_pcgame_screenshot.jpg *License*: unknown *Contributors*: Bkell, Doidimais Brasil, Melesse, N. Harmonik, SharkD, Tarinth, 1 anonymous edits

File:BalanceOfPower.png *Source*: http://en.wikipedia.org/w/index.php?title=File:BalanceOfPower.png *License*: unknown *Contributors*: Crimsonfox, McLoaf, N. Harmonik, SharkD, Tarinth

File:PoliticalMachine.jpg *Source*: http://en.wikipedia.org/w/index.php?title=File:PoliticalMachine.jpg *License*: unknown *Contributors*: Crimsonfox, Flarkins, Iridescent, SharkD, Skier Dude, Tarinth

File:Maturecity.jpg *Source*: http://en.wikipedia.org/w/index.php?title=File:Maturecity.jpg *License*: unknown *Contributors*: User:BesigedB

MIX
Papier aus verantwortungsvollen Quellen
Paper from responsible sources
FSC® C105338

Printed by Books on Demand GmbH, Norderstedt / Germany